THE SINGULARITY DUAL-ASPECT EMERGENCE (SDAE) CANON

BOOK I

COHERENCE ARCHITECTURE

An Architecture of Human Understanding of Reality

First Foundational Edition

ASTON GUSTAVOUS FARQUHARSON

NuSpecies Press
Pawling, New York, USA

Coherence Architecture:

An Architecture of Human Understanding of Reality

First Foundational Edition

ISBN 979-8-9966805-2-8 (Paperback)

ISBN 979-8-9966805-3-5 (Hardcover)

ISBN 979-8-9966805-1-1 (eBook)

Coherence Architecture is Book I of *The Singularity Dual-Aspect Emergence (SDAE) Canon*. While it stands independently as an introduction to the Architecture of Human Understanding, it also serves as the constitutional framework through which the companion SDAE volumes may be understood.

Published by NuSpecies Press

Pawling, New York, USA

Contents

FIRST FOUNDATIONAL EDITION

Every enduring architecture begins with a foundation.

This volume is offered as the First Foundational Edition of Coherence Architecture. It is not presented as the final expression of an idea, but as the first carefully assembled foundation upon which future generations may continue to build.

Reality continues to teach.

Science continues to discover.

Mathematics continues to reveal deeper relationships.

Human understanding continues to mature.

For that reason, this Architecture is intentionally open to refinement. New discoveries should strengthen it. Better mathematics should clarify it. Improved scientific understanding should deepen it. Every generation is invited to inherit these foundations with gratitude and to leave them stronger than they were received.

The measure of this work will never be whether it remained unchanged.

Its measure will be whether it remained faithful to reality.

THE BUILDER'S OATH

Before entering Reality's Cathedral, I willingly accept the discipline required of every honest learner.

I will honor reality before preference.

I will distinguish observation from interpretation.

I will distinguish empirical science from architectural recognition.

I will distinguish mathematical description from metaphor.

I will welcome correction wherever stronger evidence appears.

I will not diminish science in order to elevate philosophy.

I will not diminish philosophy in order to elevate science.

I will not confuse lived experience with empirical demonstration.

I will allow mystery to remain where evidence has not yet spoken.

I will remember that reality always remains greater than every explanation humanity constructs.

I will receive the discoveries of those who came before me with gratitude.

I will honor all those who have faithfully enlarged humanity's understanding of reality—mathematicians, scientists, physicians, engineers, educators, philosophers, historians, caregivers, artisans, and countless others whose lives became another stone in Reality's Cathedral.

I will recognize that every discipline represented within this Cathedral contributes another stone to humanity's continuing understanding.

I enter this Cathedral not as its owner,
but as one more student of reality.

May I leave Reality's Cathedral stronger than I found it.

A Final Word Before Entering

No generation begins at the beginning.

Every generation inherits.

Every generation learns.

Every generation contributes.

Every generation leaves another stone for those who follow.

If these pages encourage even one reader to become more faithful to reality,

more grateful for science,

more disciplined in thought,

more humble before mystery,

and more responsible in how they participate within the world they continue to discover,

then this Architecture will have served its purpose.

The Cathedral remains open.

Enter with curiosity.

Walk with humility.

May I become another faithful builder of Reality's Cathedral.

FOUNDATIONS OF REALITY'S CATHEDRAL

Before Entering the Cathedral

Every cathedral rests upon foundations laid by many hands.

No single architect discovers reality.

No single scientist explains it completely.

No single equation describes it exhaustively.

No single generation possesses the final word.

Reality has always been greater than every explanation humanity constructs.

This Architecture therefore begins with gratitude.

It stands upon the discoveries of countless scientists, mathematicians, physicians, engineers, philosophers, educators, historians, and ordinary people who devoted their lives to understanding reality more faithfully.

Every discovery became another stone.

Every correction strengthened the foundation.

Every unanswered question opened another doorway.

Every generation inherited more than it created.

The Architecture claims no ownership of these discoveries.

It receives them with gratitude.

It does not ask the reader to abandon the great disciplines of human inquiry.

It invites the reader to see how those disciplines belong together within one reality while preserving the integrity of each.

Reality remains the Cathedral.

Humanity remains its student.

The Three Foundational Pillars

The Architecture rests upon three enduring pillars.

First Pillar

Reality

Reality precedes every explanation.

Reality does not require belief in order to exist.

It remains humanity's first teacher.

Every discipline represented within this work ultimately derives its authority from its faithfulness to reality itself.

Second Pillar

Empirical Science

Science represents humanity's most disciplined method for learning from reality.

Observation.

Measurement.

Experimentation.

Mathematical description.

Independent verification.

Continual correction.

Science is honored throughout this work not because it possesses the final word,

but because it continually seeks greater faithfulness to reality.

Every scientific discipline represented here is received with gratitude.

Third Pillar

Mathematics

Mathematics provides humanity's most precise languages for describing relationships.

Geometry.

Calculus.

Topology.
Probability.
Information theory.
Control theory.
Network theory.
Differential equations.
These languages do not replace reality.
They illuminate relationships already present within it.
The Architecture therefore receives mathematics as one of humanity's greatest inheritances.

The Great Builders

Every generation contributes another stone to Reality's Cathedral.
Among those whose work forms part of humanity's scientific inheritance are:

Physics

Isaac Newton
James Clerk Maxwell
Albert Einstein
Max Planck
Werner Heisenberg
Erwin Schrödinger
Paul Dirac
Richard Feynman
Roger Penrose
Lee Smolin

Mathematics

Euclid
David Hilbert
Kurt Gödel
John von Neumann
Claude Shannon
Benoit Mandelbrot

Biology

Charles Darwin
Theodosius Dobzhansky
Ernst Mayr
Lynn Margulis
François Jacob
Jacques Monod
Francis Crick
Sydney Brenner
Paul Nurse

Neuroscience and Cognition

Donald Hebb
Gerald Edelman
Antonio Damasio
Giulio Tononi
Karl Friston

Systems and Complexity

Ludwig von Bertalanffy
W. Ross Ashby
Ilya Prigogine
Stuart Kauffman
John Holland
Donella Meadows

Medicine

The countless physicians,
nurses,
caregivers,
researchers,
and patients
whose shared efforts continue enlarging humanity's understanding of health,
healing,

adaptation,
and dignity.

Humanity

Every teacher.
Every parent.
Every mentor.
Every child asking why.
Every honest student.
Every careful observer.
Every person who chose truth over certainty.
Reality has always had more builders than history remembers.

The Architecture's Place

This work does not seek to replace physics.
It does not seek to replace biology.
It does not seek to replace medicine.
It does not seek to replace mathematics.
It does not seek to replace philosophy.
It does not seek to replace faith.
It seeks something different.
It asks one constitutional question.
What recurring organizational principles become visible when the discoveries of these disciplines are understood together while honoring the integrity of each?
That question belongs to Coherence Architecture.
Nothing more.
Nothing less.

The Builder's Promise

Before entering the Cathedral,
the reader is invited to make one quiet promise.
Love reality more than certainty.
Honor evidence before preference.
Receive before interpreting.
Learn before teaching.
Recognize before concluding.

Remain grateful to those who laid the stones before you.
Then add your own stone with humility,
knowing that future generations will almost certainly build farther than we have.
Reality's Cathedral has never belonged to one person,
one discipline,
or one generation.
It belongs to reality itself.
We are privileged simply to walk through it together.

WHY THIS ARCHITECTURE EXISTS

Humanity possesses extraordinary knowledge.

Physics has revealed the behavior of matter, energy, spacetime, and gravity.

Chemistry has revealed the organization of atoms, molecules, and reactions.

Biology has revealed the astonishing complexity of living systems.

Medicine has deepened our understanding of health, disease, repair, and adaptation.

Mathematics has provided humanity with precise languages through which relationships, structures, patterns, and transformations may be faithfully described.

Every discipline has enlarged humanity's inheritance.

Yet every human being lives within one reality.

The challenge before us is therefore no longer simply the accumulation of knowledge.

The challenge is the organization of understanding.

Coherence Architecture was developed in response to that challenge.

It is not proposed as another scientific discipline.

It is not offered as another philosophy.

It is not presented as a replacement for religion, mathematics, or empirical science.

It is an educational architecture.

Its purpose is to help readers recognize how the truths faithfully discovered across many disciplines belong within one coherent reality while preserving the integrity of every discipline that contributed to that understanding.

The Architecture therefore begins not with conclusions.

It begins with reality.

THE CONSTITUTIONAL PRINCIPLES

A Constitutional Declaration

Every enduring architecture is governed by principles that remain constant even as individual rooms, ideas, and discoveries continue to grow.

The purpose of this Constitution is not to restrict inquiry.

Its purpose is to preserve intellectual integrity.

The principles that follow govern every chapter of this work.

They establish the order through which empirical science, mathematics, organizational recognition, and human reflection participate together while honoring the integrity of each.

These principles are therefore offered not as immutable dogma, but as disciplined commitments to learning faithfully from reality.

Constitutional Principle I

Reality Precedes Explanation

Reality exists independently of every explanation humanity constructs.

Science, mathematics, philosophy, theology, and every human discipline ultimately remain accountable to reality itself.

Reality therefore remains humanity's first teacher.

Every explanation earns its place only to the degree that it becomes increasingly faithful to reality.

Constitutional Principle II

The Architecture Inherits Before It Interprets

Coherence Architecture introduces no competing scientific discipline.

It inherits the discoveries of empirical science.

It inherits the languages of mathematics.

Only afterward does it ask:

What recurring organizational principles become visible across those discoveries?

Inheritance therefore always precedes interpretation.

Constitutional Principle III

Scientific Integrity

Every empirical claim belongs to its proper scientific discipline.

Physics remains physics.

Chemistry remains chemistry.

Biology remains biology.

Medicine remains medicine.

Neuroscience remains neuroscience.

The Architecture introduces no new empirical mechanisms.

Its contribution is organizational recognition.

Science discovers.

The Architecture recognizes.

Constitutional Principle IV

Mathematical Integrity

Mathematics faithfully describes measurable relationships.

Architectural equations summarize recurring organizational recognitions.

They do not replace empirical equations.

Throughout this work every symbolic equation follows one constitutional order.

Empirical Foundation.

↓

Mathematical Language.

↓

Architectural Recognition.

The order never changes.

Constitutional Principle V

Organizational Recognition

The Architecture asks one recurring constitutional question throughout this work.

What recurring organizational principles become visible when reality is faithfully understood through many scientific disciplines while preserving the integrity of each?

This question defines the work.

It neither diminishes the sciences nor attempts to unify them into one discipline.

It seeks only to recognize organizational continuity across them.

Constitutional Principle VI

Coherence

Coherence is not perfection.

It is not immobility.

It is not symmetry alone.

Coherence is the continued preservation of organized relationships sufficient for identity,

function,

adaptation,

learning,

and meaningful participation within reality.

The Architecture therefore recognizes coherence as an organizational achievement rather than a static condition.

Constitutional Principle VII

Human Becoming

Reality teaches.

Science discovers.

Mathematics describes.

The Architecture recognizes.

Wisdom participates.

This progression governs the educational journey of this work.

Every chapter represents another stage in humanity's continuing participation within reality.

Constitutional Principle VIII

Intellectual Humility

Reality always remains greater than every explanation humanity constructs.

Every generation therefore inherits the responsibility to:

strengthen understanding,

correct error,

honor those who came before,

remain teachable,

and leave Reality's Cathedral stronger than it was received.

No explanation is final.

Learning continues.

Reality remains the measure.

Constitutional Principle IX

The Unity of the Cathedral

Reality is one.

Human disciplines are many.

The diversity of scientific disciplines reflects the richness of reality rather than its fragmentation.

The purpose of Coherence Architecture is therefore not to erase disciplinary distinctions.

Its purpose is to help readers recognize how those disciplines participate within one coherent reality while preserving the integrity of each.

The Cathedral belongs to Reality.

Every discipline contributes another room.

Every discovery contributes another stone.

Every generation enlarges humanity's understanding.

Closing Declaration

These Constitutional Principles govern every page that follows.

They ask the reader to enter Reality's Cathedral with intellectual humility,
scientific integrity,
mathematical discipline,
gratitude toward previous generations,
and hope for those yet to come.

Every chapter that follows simply illustrates these principles.

The Cathedral itself remains Reality.

We remain its students.

REALITY IS HUMANITY'S FIRST TEACHER

Before there were civilizations, there was reality.

Before there were philosophies, there was reality.

Before there were religions, there was reality.

Before there were scientific theories, there was reality.

Reality neither required humanity's permission nor depended upon humanity's understanding.

It simply existed.

Generation after generation, human beings gradually learned to observe its patterns.

Observation became measurement.

Measurement became mathematics.

Mathematics became prediction.

Prediction became deeper understanding.

Science emerged as humanity's most disciplined method of learning from reality.

The Architecture begins with gratitude for that inheritance.

It asks the reader to begin where every honest investigation begins:

Not with belief.

Not with ideology.

Not with inherited conclusions.

But with reality itself.

Everything else follows.

HOW TO READ THIS BOOK

This book follows one constitutional order throughout.

Every chapter proceeds through the same educational rhythm.

Reality
↓
Empirical Observation
↓
Scientific Understanding
↓
Mathematical Description
↓
Organizational Pattern
↓
Coherence Architecture
↓
Human Reflection
↓
Application

This order is intentional.

Reality precedes explanation.

Observation precedes interpretation.

Evidence precedes conclusion.

Science precedes synthesis.

Mathematics provides precision.

Coherence Architecture organizes understanding.

Reflection asks how that understanding may be lived wisely.

For this reason, readers are encouraged to move through the book in sequence.

Every chapter lays another stone.

Every chapter prepares the foundation for the one that follows.

Nothing essential is assumed.

Everything is earned.

Chapter One

Reality Before Explanation

"Reality does not ask to be believed. It asks to be discovered."

Guiding Question

What is reality, and why must every disciplined search for understanding begin with reality itself rather than with our explanations of it?

Every civilization has sought to answer profound questions about existence. Every scientific discipline has developed increasingly sophisticated methods for investigating the universe. Every philosophy has attempted to organize human understanding. Yet before any explanation can be offered, one fact remains unchanged:

Reality exists.

This chapter establishes the constitutional foundation upon which the entire Architecture is built. Before we explain reality, we must first acknowledge its independent existence. Every subsequent chapter inherits this discipline.

Scientific Foundations

This chapter stands upon the inheritance of empirical science.

Its principal foundations include:

- Astronomy and Cosmology
- Physics
- Geology
- Evolutionary Biology
- Philosophy of Science

These disciplines investigate reality through observation, measurement, experimentation, mathematical description, independent verification, and continual refinement.

Coherence Architecture neither replaces nor revises these disciplines.

It begins by receiving them with gratitude.

Their discoveries form the constitutional foundation upon which this Architecture is built.

Mathematical Foundations

Scientific understanding becomes increasingly precise because it is accompanied by mathematics.

This chapter therefore inherits the mathematical languages through which reality becomes measurable and its relationships increasingly intelligible.

Among them are:

- Measurement
- Geometry
- Calculus
- Probability
- Mathematical Modeling

Mathematics does not create reality.

Neither does it compete with observation.

It extends observation.

It enables humanity not only to recognize patterns, but to test them, quantify them, compare them, and predict them with extraordinary precision.

In this sense, mathematics becomes one of science's greatest partners in humanity's continuing conversation with reality.

Architectural Recognition

Only after empirical science and mathematics have spoken does the Architecture introduce its first constitutional recognition.

Reality precedes every explanation.

Every scientific theory.

Every mathematical description.

Every philosophical system.

Every civilization.

Every generation.

Represents humanity's continuing effort to become increasingly faithful to a reality that existed long before humanity learned how to describe it.

This recognition is not proposed as another scientific theory.

Nor does it replace empirical investigation.

It simply establishes the constitutional order that governs every chapter that follows.

Reality first.

Explanation second.

Reality Before Explanation

"Reality does not ask to be believed. It asks to be discovered."

Long before humanity possessed written language, mathematical symbols, or scientific instruments, reality already existed.

Stars ignited.

Galaxies evolved.

Mountains rose.

Rivers carved valleys.

Atoms combined into molecules.

Life emerged.

Reality neither awaited human understanding nor depended upon human opinion.

It unfolded according to principles humanity would gradually discover through observation, measurement, mathematics, and disciplined inquiry.

This simple observation becomes the first constitutional principle of Coherence Architecture.

Reality precedes every explanation.

Every scientific theory, every mathematical equation, every philosophical reflection, and every spiritual tradition represents humanity's continuing effort to understand a reality that existed long before humanity learned how to describe it.

The Architecture therefore begins with an act of intellectual humility.

Before we attempt to explain reality, we must first learn to observe it faithfully.

For only reality possesses the authority to judge the adequacy of our explanations.

Everything else remains provisional.

Reality Exists Before Understanding

Human understanding has never created reality.

It has only become progressively more faithful to it.

The Earth revolved around the Sun long before astronomy demonstrated it.

Gravity shaped galaxies long before mathematics described gravitational attraction.

Evolution unfolded across immense timescales before biology recognized the mechanisms through which life diversified.

Reality does not become true because humanity finally understands it.

Human understanding becomes more faithful as it increasingly conforms to reality.

This distinction is fundamental.

The purpose of science is not to invent reality.

Its purpose is to discover reality more faithfully than the generation before.

Every generation therefore inherits two responsibilities.

To understand reality more faithfully.

And to leave that understanding better organized for those who follow.

The Scientific Discipline

Modern science represents one of humanity's greatest achievements because it developed a disciplined method for learning from reality rather than merely speculating about it.

Scientific understanding advances through disciplined practice:

- Observation
- Measurement
- Hypothesis
- Experimentation
- Mathematical Description
- Independent Verification
- Continual Correction

These disciplines transformed human civilization.

Science does not become powerful because it claims certainty.

Science becomes powerful because it continually welcomes correction.

Every better telescope reveals more of the universe.

Every better microscope reveals more of life.

Every better instrument enlarges humanity's conversation with reality.

For this reason, Coherence Architecture begins by expressing gratitude toward empirical science.

Science does not diminish wonder.

Science enlarges wonder.

The more faithfully humanity observes reality, the more remarkable reality becomes.

Mathematics: The Language of Relationship

Observation alone cannot fully organize understanding.

Patterns must also be described.

Mathematics provides humanity with one of its most precise languages for accomplishing that task.

Geometry reveals form.

Topology reveals continuity.

Graph theory reveals relationships.

Probability reveals uncertainty.

Calculus reveals change.

Information theory reveals communication.

Across every scientific discipline, mathematics performs the same essential function.

It faithfully describes relationships already present within reality.

Mathematics does not compete with observation.

It extends observation.

It enables humanity not only to recognize patterns, but to test them, quantify them, compare them, and predict them with extraordinary precision.

The Architecture therefore inherits mathematics with gratitude.

It neither replaces mathematics nor diminishes it.

Instead, it asks a different question.

What recurring organizational patterns become visible when mathematics faithfully describes many different domains of reality?

That question belongs neither exclusively to physics, biology, chemistry, nor mathematics.

It belongs to the organization of understanding itself.

From Observation to Organization

The purpose of this Architecture is not to replace the sciences.

Physics investigates matter, energy, motion, and spacetime.

Chemistry investigates atoms, molecules, and reactions.

Biology investigates living systems.

Medicine investigates health, disease, adaptation, and repair.

Each discipline possesses its own methods, evidence, mathematical frameworks, and standards of proof.

Having inherited the discoveries of empirical science and the descriptive power of mathematics, the Architecture now asks another question.

What recurring organizational principles become visible across these disciplines while fully respecting the integrity of each?

The Architecture does not claim that physics becomes biology.

Nor that biology becomes psychology.

Nor that mathematics becomes philosophy.

Rather, it asks whether reality, faithfully investigated through many disciplines, exhibits recurring organizational patterns worthy of recognition.

Reality is one.

Our disciplines are many.

The Architecture seeks not to erase those distinctions, but to help the reader understand how they belong together.

The First Principle

Every enduring structure begins with a foundation.

No architect begins by constructing the highest tower.

Every master builder begins with the first stone.

Human understanding follows the same discipline.

This Architecture therefore begins not with consciousness.

Not with civilization.

Not with spirituality.

Not even with coherence.

It begins where every honest investigation must begin.

With reality.

This constitutional order remains unchanged throughout the Architecture.

Reality first.

Explanation second.

From reality emerge relationships.

From relationships emerge distinction.

From distinction emerge boundaries.

From boundaries emerge vulnerability.

From vulnerability emerge adaptation.

From adaptation emerge awareness, learning, memory, knowledge, meaning, and eventually coherence itself.

Nothing essential will be assumed.

Every conclusion will be earned.

Every chapter will rest upon foundations already established.

The reader is invited to walk this path patiently.

The Architecture unfolds in the same direction that organized reality appears to unfold.

The Architect's Reflection

Human beings naturally seek answers.

Reality often teaches something even more valuable.

It teaches better questions.

Many of humanity's greatest discoveries began when someone refused to believe that the familiar explanation was sufficient.

A falling apple became a question.

The motion of planets became a question.

The behavior of light became a question.

The structure of living cells became a question.

The inheritance of traits became a question.

Questions opened the door through which deeper understanding entered civilization.

Reality rewards disciplined curiosity.

Every generation inherits mysteries.

Every generation transforms some of those mysteries into understanding.

Every generation also discovers new mysteries beyond the horizon.

The conversation between humanity and reality is therefore never finished.

This Architecture is offered as one contribution to that continuing conversation.

Quiet Reflection

Before continuing, pause.

Look around you.
Choose one ordinary object.
A leaf.
A stone.
A cup.
A tree beyond the window.
For a moment, resist the urge to explain it.
Simply acknowledge that it exists independently of your understanding.
This simple act is the beginning of intellectual humility.
Reality is not waiting for our explanations.
Our explanations are continually learning to become more faithful to reality.

Crossing the Next Arch

Reality has now been established as the first teacher.
Another question naturally appears.
Does anything within reality exist entirely alone?
Stars exist within galaxies.
Atoms exist within molecules.
Cells exist within organisms.
Organisms exist within ecosystems.
Human beings exist within families, communities, civilizations, and histories.
Reality therefore presents itself not merely as isolated objects.
It presents itself as relationships.
Understanding deepens when relationships become visible.
The next chapter begins there.
For only when relationships become visible does the Architecture begin revealing its second arch.

Cathedral Thread

Constitutional Principle Established
Reality precedes explanation.
Everything that follows in this Architecture inherits that discipline.
Science observes.
Mathematics describes.
The Architecture recognizes.
Wisdom responds.
The next chapter asks the first great relational question:

If reality is one, how are its relationships organized?
The Cathedral now opens its second chamber.

Chapter Two

Relation Before Isolation

Guiding Question

Why are relationships fundamental to reality, and why does understanding deepen only as those relationships become visible?

Every scientific discipline begins by observing things.

The Architecture now asks the next constitutional question.

How do those things belong together?

Reality does not merely present isolated objects for observation.

It presents relationships.

This chapter establishes the second constitutional principle of the Architecture.

Reality becomes increasingly intelligible as its relationships become increasingly visible.

Scientific Foundations

This chapter stands upon the inheritance of empirical science.

Its principal foundations include:

- General Relativity
- Electromagnetism
- Chemistry
- Molecular Biology
- Ecology
- Physiology

- Neuroscience
- Systems Science

Each discipline reveals that organization emerges through interaction rather than isolation.

Mass relates to spacetime.

Charges relate through electromagnetic fields.

Atoms relate through chemical bonding.

Cells relate within tissues.

Organs relate within organisms.

Organisms relate within ecosystems.

Brains function through neural networks.

Civilizations endure through relationships among people, institutions, and knowledge.

The mechanisms differ.

The recurring observation remains.

Reality continually presents itself through relationships.

Mathematical Foundations

Mathematics provides several complementary languages through which relationships become increasingly intelligible.

Among them are:

- Geometry
- Topology
- Graph Theory
- Network Theory
- Information Theory
- Dynamical Systems

Each illuminates different aspects of connectedness.

Geometry reveals spatial relationships.

Topology reveals continuity.

Graph theory reveals networks.

Information theory reveals communication.

Systems mathematics reveals interaction.
None alone explains reality.
Together they make relationships increasingly visible.
The Architecture inherits these mathematical languages with gratitude.

Architectural Recognition

Only after empirical science and mathematics have spoken does the Architecture introduce its second constitutional recognition.

Nothing within reality becomes fully intelligible until its relationships become visible.

The Architecture does not claim that relationships replace objects.

Nor does it suggest that every discipline studies relationships in the same way.

Rather, it recognizes that understanding repeatedly deepens as relationships become visible across every scale presently known to science.

Reality remains one.

Relationships reveal its organization.

Relation Before Isolation

"Nothing within reality becomes fully intelligible until its relationships become visible."

The first chapter established the constitutional foundation of this Architecture.

Reality exists independently of our understanding.

The second constitutional principle follows naturally.

Nothing within reality exists entirely alone.

Every observable phenomenon participates in relationships.

Stars exist within galaxies.

Galaxies participate within larger gravitational structures.

Atoms interact through electromagnetic forces.

Molecules emerge through chemical bonding.

Cells participate within tissues.

Organs function within organisms.

Organisms participate within ecosystems.

Human beings live within families, communities, civilizations, histories, and cultures.

Reality therefore presents itself not merely as isolated entities.

It presents itself as organized relationships.

Objects attract our attention.

Relationships explain their participation.

This observation becomes the second constitutional principle of Coherence Architecture.

Understanding deepens as relationships become increasingly visible.

The Relational Universe

For centuries, human understanding advanced by examining reality through increasingly specialized disciplines.

Astronomy investigated the heavens.

Physics investigated matter, energy, motion, and spacetime.

Chemistry investigated atoms and molecules.

Biology investigated living systems.

Medicine investigated health and disease.

Mathematics revealed relationships through precise description.

Each discipline enlarged humanity's inheritance.

Each revealed truths previously hidden.

Yet specialization also introduced an understandable challenge.

The deeper humanity understood individual domains, the more difficult it sometimes became to recognize how those domains belonged together.

The Architecture does not diminish specialization.

It depends upon it.

Without rigorous disciplines there can be no trustworthy synthesis.

The purpose of Coherence Architecture is therefore not to reduce many sciences into one.

Its purpose is to help readers recognize how the discoveries of many disciplines participate within one reality while preserving the integrity of every discipline that contributed to that understanding.

Empirical Science Reveals Relationships

Modern science increasingly demonstrates that relationships are not merely secondary characteristics of reality.

They are among its most remarkable organizing principles.

Gravity relates mass to spacetime.

Electromagnetism relates charged particles through fields.

Chemical bonding relates atoms into molecules.

DNA relates one generation of life to the next through inherited biological information.

Ecology reveals relationships among organisms and their environments.

Neuroscience investigates coordinated neural networks rather than isolated neurons.

Physiology reveals that health depends upon the coordinated function of many organ systems rather than any single organ acting alone.

Across the sciences, the mechanisms differ profoundly.

Yet the lesson quietly repeats.

Understanding grows as relationships become visible.

The Architecture inherits that lesson with gratitude.

Mathematics: The Language of Connection

Mathematics provides several complementary languages through which relationships may be described with extraordinary precision.

Geometry reveals spatial relationships.

Topology studies continuity despite transformation.

Graph theory reveals networks.

Information theory reveals communication.

Systems theory reveals feedback.

Dynamical systems describe continual interaction.

Each mathematical language illuminates different forms of relationship.

No single mathematical framework exhausts reality.

Together they reveal something remarkable.

Reality becomes increasingly intelligible through relationship.

The Architecture receives these mathematical achievements exactly where they belong.

Not as replacements for observation.

But as humanity's most refined language for describing the relationships observation continually reveals.

Organization Emerges Through Relationship

Relationships do more than connect existing entities.

Relationships organize possibility.

Hydrogen and oxygen participate in relationships that give rise to water.

Individual cells participate in relationships that become tissues.

Tissues participate in relationships that become organs.

Organs participate in relationships that become living organisms.

Individuals participate in relationships that become families, communities, institutions, and civilizations.

Organization therefore emerges not from isolation, but from coordinated interaction.

The Architecture recognizes this recurring pattern across every scale presently known to science.

The mechanisms differ.

The organizational principle remains.

Increasingly organized systems arise through increasingly organized relationships.

The Principle of Relational Becoming

Every enduring system participates simultaneously in two complementary realities.

Its own internal organization.

Its relationships with everything beyond itself.

A living cell survives through continual exchange with its environment.

A star exists within the gravitational relationships of its galaxy.

An ecosystem survives through continual interaction among countless living organisms.

A civilization survives through education, communication, cooperation, and inherited knowledge.

Isolation rarely strengthens organization.

Meaningful relationship sustains it.

The Architecture therefore proposes its second constitutional principle.

Nothing meaningful becomes fully intelligible in isolation.

Every explanation depends upon context.

Every context consists of relationships.

Understanding deepens as relationships become increasingly visible.

Human Relationships

The same constitutional principle extends naturally into human life.

No child becomes fully human in isolation.

Language develops through relationship.

Learning develops through relationship.

Character develops through relationship.

Scientific understanding develops through relationship.

Culture develops through relationship.

Civilization itself becomes possible because understanding passes from one generation to another.

Every teacher.

Every scientist.

Every physician.

Every engineer.

Every parent.

Every caregiver.

Participates in one of humanity's oldest forms of relational inheritance.

Knowledge survives because relationships preserve it.

Education therefore becomes one of civilization's greatest relational achievements.

The Architect's Reflection

For many years I believed that understanding consisted primarily of acquiring more information.

Reality gradually taught another lesson.

People rarely suffer because they lack isolated facts.

More often they suffer because they cannot see how the parts of their lives belong together.

Health influences purpose.

Purpose influences hope.

Hope strengthens perseverance.

Relationships shape identity.

Identity shapes responsibility.

The deepest understanding often arrives not when another fact is discovered, but when previously separate truths suddenly recognize one another.

Perhaps wisdom begins when relationships become more visible than isolated facts.

Quiet Reflection

Pause before continuing.

Choose one ordinary object nearby.

A tree.

A stone.

A cup.

Now ask three questions.

What relationships made this possible?

What relationships continue sustaining it?

What relationships will remain after it changes?

Remain with those questions.

Learning to perceive relationships is itself one of the first disciplines of coherent understanding.

Crossing the Next Arch

Relationships alone do not produce identity.

For one relationship to be recognized as distinct from another,

something must become distinguishable.

Difference must emerge.

Identity must appear.

The next chapter explores one of reality's most profound organizational transitions.

Differentiation.

Only when distinction becomes possible can increasingly complex forms of organization begin.

Only then can reality become capable of history.

Only then does becoming truly begin.

Cathedral Thread

Constitutional Principle Established

Nothing within reality becomes fully intelligible until its relationships become visible.

Chapter One established that Reality precedes explanation.

Chapter Two establishes that Understanding deepens through relationship.

The next chapter asks another constitutional question.

How does relationship become identity?

The Cathedral now opens its third chamber.

Chapter Three

Differentiation

The Birth of Identity

"The universe does not become richer by remaining undifferentiated. It becomes richer because distinction makes organization possible."

Guiding Question

How does reality move from relationship to distinguishable identity, and why is differentiation one of the fundamental conditions for increasingly organized existence?

The previous chapter established that understanding deepens through relationships.

Yet relationships alone cannot produce recognizable organization.

For one relationship to be distinguished from another,

difference must emerge.

Identity must appear.

This chapter explores one of reality's most profound organizational transitions.

Differentiation.

Without differentiation there can be no stars.

No atoms.

No chemistry.

No biology.

No observer.

No history.

Differentiation therefore becomes one of the great conditions that makes organized becoming possible.

Scientific Foundations

This chapter stands upon the inheritance of empirical science.

Its principal foundations include:

- Cosmology
- Particle Physics
- Atomic Physics
- Chemistry
- Evolutionary Biology
- Developmental Biology
- Embryology
- Neuroscience

Across these disciplines, increasingly sophisticated organization emerges through successive stages of differentiation.

Particles become distinguishable.

Atoms become distinguishable.

Elements become distinguishable.

Cells become specialized.

Organs become differentiated.

Brains develop functional regions.

Species diversify.

The mechanisms differ profoundly.

The recurring observation remains.

Increasing organization is accompanied by increasing differentiation.

The Architecture inherits that observation with gratitude.

Mathematical Foundations

Mathematics provides several complementary languages through which differentiation becomes increasingly intelligible.

Among them are:

- Geometry
- Topology
- Group Theory
- Symmetry

- Graph Theory
- Fractal Mathematics
- Information Theory

Geometry distinguishes form.

Topology preserves continuity while permitting transformation.

Group theory investigates symmetry and symmetry breaking.

Fractal mathematics illustrates recursive differentiation across scales.

Information theory distinguishes meaningful signal from background uncertainty.

Each mathematical language contributes another perspective.

Together they reveal that differentiation is not disorder.

It is organization becoming increasingly expressive.

Architectural Recognition

Only after empirical science and mathematics have spoken does the Architecture introduce its third constitutional recognition.

Organization requires distinction.

Relationships alone do not generate identity.

Identity emerges only when reality becomes capable of differentiation.

The Architecture does not propose that every scientific mechanism of differentiation is identical.

Rather, it recognizes that increasingly organized reality repeatedly becomes increasingly distinguishable.

Reality remains one.

Differentiation reveals its richness.

Differentiation

The Birth of Identity

Reality exists.

Reality relates.

Yet relationships alone do not produce distinguishable existence.

For anything to become recognizable,

reality must permit differentiation.
Without differentiation,
no region of the universe could become distinguishable from another.
Without differentiation,
no particles could possess unique identities.
Without differentiation,
no elements could emerge.
Without differentiation,
no chemistry could unfold.
Without differentiation,
no living organisms could become specialized.
Without differentiation,
there could be no observer,
because there would be no distinction between observer and observed.
Differentiation therefore is not merely another event within cosmic history.
It is one of the fundamental conditions that makes history itself possible.

The Architecture recognizes differentiation as one of reality's first great organizational achievements.

Organization requires distinction.
Distinction permits identity.
Identity prepares becoming.

The Scientific Inheritance

Across every major scientific discipline, differentiation appears as one of reality's recurring organizational principles.

Modern cosmology describes an evolving universe in which increasingly complex structures emerged through successive stages of differentiation.

Particle physics investigates distinguishable particles and their interactions.

Atomic physics investigates distinguishable elements.

Chemistry investigates increasingly differentiated compounds.

Biology investigates differentiated cells, tissues, organs, organisms, and species.

Embryology studies how one fertilized cell gradually differentiates into the extraordinary diversity of tissues that constitute a living body.

Evolutionary biology studies the continual differentiation of populations across immense spans of time.

Neuroscience reveals increasing differentiation within the nervous system, where specialized regions perform distinct yet coordinated functions.

Throughout empirical science, the mechanisms differ profoundly.

Yet one recurring observation quietly appears.

Increasing organization is accompanied by increasing differentiation.

The Architecture neither replaces nor simplifies these scientific explanations.

It inherits them.

Only afterward does it ask what broader organizational principle they collectively reveal.

Mathematics and the Emergence of Difference

Mathematics provides several complementary languages through which differentiation becomes increasingly intelligible.

Geometry distinguishes form.

Topology preserves continuity while allowing transformation.

Group theory investigates symmetry and symmetry breaking.

Graph theory describes increasingly differentiated networks.

Fractal mathematics illustrates recursive differentiation across multiple scales.

Information theory distinguishes meaningful signal from surrounding uncertainty.

Each mathematical discipline contributes a different perspective.

Together they illuminate an important truth.

Difference is not disorder.

Difference is organization becoming increasingly expressive.

The Architecture therefore receives mathematics not merely as calculation,

but as one of humanity's greatest instruments for recognizing the structure of differentiated reality.

Symmetry and the Birth of Complexity

One of the deepest insights emerging from modern physics is that many forms of organization become possible through the breaking of prior symmetries.

Perfect symmetry possesses remarkable elegance.

Yet perfect symmetry contains little distinction.

When symmetry breaks,

new possibilities emerge.

Different particles.

Different interactions.

Different structures.

Different histories.

Modern cosmology and particle physics continue investigating these extraordinary transitions.

The Architecture does not attempt to replace these investigations.

Instead, it recognizes that symmetry breaking illustrates a broader organizational principle.

Increasing complexity often becomes possible through increasing differentiation.

The universe does not become poorer because distinction appears.

It becomes richer because possibility expands.

Identity Through Differentiation

Differentiation gives rise to identity.

Identity should never be confused with isolation.

Identity makes meaningful relationship possible.

Every star possesses its own evolutionary history.

Every planet follows its own trajectory.

Every organism develops its own biological history.

Every civilization follows its own cultural journey.

Every human being carries a story unlike any other.

Identity therefore represents one of reality's most remarkable achievements.

Not because identity separates.

Because identity allows participation.

The Architecture understands identity not as separation from reality,
but as a unique way of participating within reality.

Recursive Differentiation

Reality rarely differentiates only once.

Differentiation becomes recursive.

Galaxies differentiate into stellar systems.

Stellar systems differentiate into planets.

Planets differentiate into ecosystems.

Life differentiates into organisms.

Organisms differentiate into tissues.

Brains differentiate into specialized regions.

Human cultures differentiate into languages, traditions, institutions, and bodies of knowledge.

Each new level inherits previous organization while introducing genuinely new capacities.

Differentiation therefore does more than increase complexity.

It generates history.

Every differentiated system now enters its own unique process of becoming.

The Principle of Differentiated Becoming

Every differentiated system acquires possibilities unavailable to undifferentiated existence.

Once identity appears,
interaction becomes richer.

Adaptation becomes possible.

Learning becomes possible.

Inheritance becomes possible.

Creativity becomes possible.

Failure becomes possible.

Renewal becomes possible.

Differentiation therefore introduces both opportunity and responsibility.

Reality no longer consists merely of existence.

Reality becomes history.

The Architecture recognizes this continuing process as Differentiated Becoming.

Every differentiated system carries forward what preceded it while contributing something genuinely new to what follows.

Human Differentiation

The same constitutional principle extends naturally into human life.

No two individuals possess identical histories.

No two minds develop through identical experiences.

Families cultivate different personalities.

Communities cultivate different gifts.

Education differentiates understanding.

Scientific disciplines differentiate methods.

Civilizations differentiate institutions.

These distinctions should not be understood as fragmentation.

Properly organized,
differentiation enriches coherence.

The healthiest societies are not those in which every individual becomes identical.

They are those in which differentiated strengths contribute toward shared flourishing.

Diversity without relationship fragments.

Relationship without differentiation stagnates.

Coherence requires both.

The Architect's Reflection

For many years I believed individuality meant becoming increasingly separate from others.

Reality gradually taught another lesson.

Every meaningful identity is formed through relationships.

Parents shape children.

Teachers shape students.

Communities shape character.

Reality shapes every life.

Differentiation therefore does not diminish belonging.

It deepens contribution.

Perhaps maturity is not becoming less connected.

Perhaps maturity is becoming increasingly capable of offering one's unique gifts in service to something greater than oneself.

Quiet Reflection

Walk through a forest.

At first you see many trees.

Remain a little longer.

Each tree possesses its own shape.

Its own scars.

Its own branches.

Its own history.

Yet together they become something none could become alone.

The forest is not diminished because its trees differ.

Its richness depends upon their differentiation.

The same is true of humanity.

The same is true of understanding.

Crossing the Next Arch

Differentiation makes identity possible.

Yet identity cannot endure without distinction.

Every differentiated system therefore acquires boundaries.

A cell develops a membrane.

A planet possesses an atmosphere.

An organism develops skin.

A civilization establishes institutions.

Boundaries preserve identity.

Yet they also introduce something entirely new.

Vulnerability.

The next chapter explores one of the deepest paradoxes in organized reality.

The very condition that allows identity to exist also creates the possibility of injury.

Without boundaries,
nothing remains itself.

With boundaries,
everything becomes capable of both persistence and loss.

Only here does the Architecture begin revealing why coherence becomes one of reality's most remarkable achievements.

Cathedral Thread

Constitutional Principle Established

Organization requires distinction. Distinction permits identity.

Chapter One established that Reality precedes explanation.

Chapter Two established that Understanding deepens through relationship.

Chapter Three establishes that Identity emerges through differentiation.

The next chapter asks another constitutional question.

How is identity preserved?

The Cathedral now opens its fourth chamber.

Chapter Four

Boundary

The Birth of Individuality

"The first boundary did not merely separate. It made identity possible."

Guiding Question

How does differentiation become enduring identity, and why are boundaries among the fundamental conditions of organized existence?

The previous chapter established that differentiation allows reality to become distinguishable.

Yet differentiation alone cannot preserve identity.

Every distinguishable system must also possess a boundary.

Without boundaries there can be no lasting distinction.

Without distinction there can be no enduring organization.

This chapter explores one of reality's most remarkable organizational achievements.

The boundary.

Far from merely separating one thing from another,
boundaries define,
protect,
regulate,
and sustain identity.

The Architecture therefore recognizes boundary as one of the great constitutional conditions of organized persistence.

Scientific Foundations

This chapter stands upon the inheritance of empirical science.

Its principal foundations include:

- Cell Biology
- Physiology

- Immunology
- Planetary Science
- Astrophysics
- Ecology
- Systems Biology

Across these disciplines,

organized systems persist because they maintain boundaries while continually interacting with their environments.

Cells possess selectively permeable membranes.

The nucleus protects genetic information.

Skin protects organisms while permitting exchange.

Atmospheres regulate planetary environments.

Stars possess internal layers that regulate energy transport.

Ecosystems exhibit recognizable boundaries while exchanging matter and energy with neighboring systems.

The mechanisms differ.

The recurring observation remains.

Organization requires boundaries.

Mathematical Foundations

Mathematics likewise provides several complementary languages through which boundaries become increasingly intelligible.

Among them are:

- Geometry
- Topology
- Boundary Value Theory
- Differential Geometry
- Graph Theory
- Information Theory

Geometry distinguishes interior from exterior.

Topology studies continuity across boundaries.

Boundary value problems investigate interactions occurring at interfaces.

Graph theory distinguishes internal and external connectivity.

Information theory recognizes interfaces through which communication becomes possible.

Together they reveal an important insight.

Boundaries are not merely edges.

They are organized regions through which relationships become regulated.

Architectural Recognition

Only after empirical science and mathematics have spoken does the Architecture introduce its fourth constitutional recognition.

Identity is preserved through organized boundaries.

Boundaries should never be confused with isolation.

Healthy boundaries regulate participation.

They preserve identity while allowing relationship.

The Architecture therefore recognizes boundaries not as barriers,
but as organized interfaces between interior coherence and exterior reality.

Boundary

The Birth of Individuality

The previous chapter established that differentiation permits identity.

Stars become distinguishable.

Atoms become distinguishable.

Living organisms become distinguishable.

Civilizations develop distinct histories.

Identity, however,
cannot endure through differentiation alone.

Every identity must possess a boundary.

Without boundaries,
there can be no meaningful distinction between one organized system and another.

Without distinction,
there can be no enduring persistence.

Boundary therefore becomes one of reality's most remarkable organizational achievements.

It does not merely separate.

It defines.

It regulates.

It preserves.

The Architecture therefore recognizes boundary as one of the first constitutional conditions through which organized existence becomes capable of enduring.

The Scientific Inheritance

Throughout empirical science, boundaries appear wherever organized systems persist.

Cells possess selectively permeable membranes that regulate the movement of molecules.

The nucleus protects genetic information while permitting carefully regulated communication with the surrounding cell.

Organs are distinguished through specialized tissues that maintain both structural integrity and functional coordination.

The skin protects the body while continually exchanging heat, moisture, gases, and sensory information with the external environment.

Planets possess atmospheres that regulate temperature, chemistry, and energy exchange.

Stars possess internal layers through which energy is generated, transported, and ultimately radiated into space.

Galaxies maintain recognizable gravitational organization while continually interacting with neighboring galaxies and the expanding universe.

Ecological communities possess identifiable boundaries while continually exchanging organisms, matter, and energy with surrounding ecosystems.

Across every discipline,

the mechanisms differ.

The recurring observation remains remarkably consistent.

Organization requires boundaries.

Boundaries preserve organization.

The Architecture receives this scientific inheritance with gratitude.

Only afterward does it ask what broader organizational principle these observations collectively reveal.

Mathematics and Boundary

Mathematics has long recognized the extraordinary significance of boundaries.

Geometry distinguishes interior from exterior.

Topology investigates continuity across changing boundaries.

Boundary value theory examines how systems behave at interfaces.

Differential geometry studies surfaces through which complex structures become intelligible.

Graph theory distinguishes internal connectivity from external connectivity.

Information theory recognizes interfaces through which communication becomes possible.

Each mathematical language reveals another aspect of organized distinction.

Boundaries are not merely limits.

They are regions where relationships become regulated.

The Architecture inherits these mathematical insights because they illuminate one of reality's deepest recurring principles.

Identity is preserved through organized boundaries.

Boundary as Organized Participation

Boundaries should never be confused with isolation.

A healthy cell does not seal itself completely from its surroundings.

A healthy organism continually exchanges oxygen, nutrients, water, energy, and information with the world.

A healthy ecosystem exchanges organisms, resources, and energy with neighboring systems.

A healthy civilization learns from others while preserving its own integrity.

Boundaries therefore perform a remarkable function.

They regulate participation.

They determine what enters.

They determine what leaves.

They preserve internal organization while permitting meaningful relationship within larger systems.

The Architecture therefore understands boundaries not as barriers,
but as organized interfaces.

They do not exist to prevent relationship.

They exist to organize relationship.

The Principle of Interior and Exterior

Every boundary establishes two complementary domains.
An interior.
An exterior.
The interior preserves organized identity.
The exterior presents opportunity.
Resources.
Novelty.
Challenge.
Uncertainty.
Neither domain alone explains persistence.
The living cell survives because it continually negotiates both.
The organism survives because it continually negotiates both.
The civilization survives because it continually negotiates both.
The same constitutional principle appears throughout organized reality.
Coherence emerges through disciplined participation between interior organization and exterior reality.
Neither complete openness nor complete isolation sustains life.
Wisdom resides in regulation.

The Birth of Vulnerability

The appearance of boundaries introduces something entirely new into the history of organized existence.
Vulnerability.
Before boundaries,
there is little to preserve.
After boundaries,
every organized system becomes capable of disruption.
Cells may be damaged.
Stars may collapse.
Species may become extinct.
Civilizations may decline.
Human beings may suffer loss.
The very condition that preserves identity simultaneously creates the possibility of injury.
This is not a flaw within reality.

It is one of reality's deepest organizational consequences.
Every identity worthy of preservation becomes vulnerable.
The Architecture therefore regards vulnerability not as failure.
It is the inevitable companion of individuality.

Boundaries Across Human Life

The same constitutional principle quietly shapes human experience.
Families possess boundaries.
Communities possess boundaries.
Cultures possess boundaries.
Scientific disciplines possess methodological boundaries.
Professional ethics establish boundaries.
Personal integrity depends upon boundaries.
Healthy boundaries do not isolate human beings.
They preserve trust.
They cultivate responsibility.
They make genuine relationship possible.
Without boundaries,
there can be little identity.
Without identity,
there can be little responsibility.
Without responsibility,
civilization itself begins to weaken.
The Architecture therefore recognizes boundaries as one of the great protectors of coherent human life.

The Architect's Reflection

Many people spend much of life attempting to eliminate boundaries.
Reality teaches another lesson.
Every enduring relationship depends upon healthy boundaries.
Parents protect children through boundaries.
Teachers cultivate learning through boundaries.
Scientific inquiry advances because disciplined methods establish intellectual boundaries.
Civilizations flourish because justice establishes lawful boundaries.
Strength therefore does not arise from the absence of boundaries.
Strength arises from boundaries wisely maintained.

Perhaps the highest boundaries are not walls.
They are commitments.
Commitments to truth.
Commitments to integrity.
Commitments to compassion.
These boundaries do not imprison us.
They allow us to become trustworthy.

Quiet Reflection

Place your hand over your heart.

Beneath your hand are countless boundaries.

Skin.

Cells.

Membranes.

Tissues.

Organs.

Every one quietly preserving life.

Your body survives not because boundaries isolate you from reality.

Your body survives because boundaries continually regulate your participation within reality.

Remain with that thought.

Perhaps every meaningful relationship in your life also depends upon boundaries that protect without separating.

Crossing the Next Arch

Every boundary preserves identity.

Every preserved identity becomes vulnerable.

The question now becomes unavoidable.

How does anything endure?

How does a star remain stable for billions of years?

How does a living cell continually repair itself?

How does a forest recover after disturbance?

How does a civilization preserve knowledge across generations?

How does a human life remain faithful to itself while continually changing?

The next chapter enters one of the deepest questions in science and in human experience.

Vulnerability.

For only when reality becomes vulnerable does adaptation become necessary. Only then does coherence begin its lifelong work.

Cathedral Thread

Constitutional Principle Established

Identity is preserved through organized boundaries.
Chapter One established:
Reality precedes explanation.
Chapter Two established:
Understanding deepens through relationship.
Chapter Three established:
Identity emerges through differentiation.
Chapter Four establishes:
Boundaries preserve identity while regulating participation.
The next chapter asks the inevitable constitutional question:
Why does every organized identity become vulnerable?
The Cathedral now opens its fifth chamber.

Chapter Five

Vulnerability

The Birthplace of Becoming

"Every boundary preserves an identity. Every boundary also creates a vulnerability."

Guiding Question

Why does every organized system become vulnerable, and why is vulnerability not the opposite of organization, but one of its deepest consequences?

The previous chapter established that boundaries preserve identity.

Every organized system capable of persistence possesses something worth preserving.

Yet that same achievement introduces an unavoidable consequence.

The moment identity appears,
the possibility of loss also appears.

This chapter explores one of reality's most profound organizational truths.

Vulnerability is not an accident of organized existence.

It is one of its inevitable consequences.

Without vulnerability,
there is nothing to preserve.

Without something to preserve,
there can be no history.

No adaptation.

No responsibility.

No becoming.

The Architecture therefore recognizes vulnerability not as weakness,
but as one of the great constitutional conditions through which increasingly organized reality continues its unfolding.

Scientific Foundations

This chapter stands upon the inheritance of empirical science.

Its principal foundations include:

- Thermodynamics
- Astrophysics
- Cell Biology
- Physiology
- Evolutionary Biology
- Ecology
- Systems Biology
- Medicine

Across these disciplines,

organized systems continually encounter forces capable of disrupting their organization.

Stars continually resist gravitational collapse through hydrostatic equilibrium.

Living cells continually repair molecular damage.

Organisms continually regulate temperature, hydration, metabolism, and immune function.

Ecosystems continually recover from disturbance.

Civilizations continually preserve knowledge despite conflict, environmental change, and cultural transformation.

The mechanisms differ profoundly.

The recurring observation remains.

Every organized system encounters continual coherence challenges.

The Architecture inherits this scientific observation with gratitude.

Mathematical Foundations

Mathematics likewise provides powerful languages through which vulnerability becomes increasingly intelligible.

Among them are:

- Thermodynamics
- Differential Equations
- Control Theory

- Network Theory
- Information Theory
- Dynamical Systems

These mathematical disciplines describe systems continually responding to disturbance while preserving recognizable organization.

Persistence is never passive.

Organization is continually maintained.

The Architecture receives these mathematical languages because they illuminate one remarkable constitutional principle.

Vulnerability is not disorder.

Vulnerability is the condition that makes organized persistence necessary.

Architectural Recognition

Only after empirical science and mathematics have spoken does the Architecture introduce its fifth constitutional recognition.

Every organized identity becomes vulnerable.

The greater the organization,
the greater the responsibility required to preserve it.

The Architecture does not regard vulnerability as failure.

It recognizes vulnerability as the inevitable consequence of identity becoming worthy of preservation.

Vulnerability

The Birthplace of Becoming

The previous chapter established that boundaries preserve identity.

A living cell exists because its membrane distinguishes its interior from the surrounding environment.

A planet exists because gravity gathers matter into organized persistence.

A star exists because its immense mass remains sufficiently organized rather than dispersing into space.

An organism exists because countless biological boundaries preserve coordinated function.

Every civilization exists because language, institutions, memory, and culture preserve recognizable identity across generations.

Boundaries make individuality possible.

Yet boundaries also introduce one of reality's deepest paradoxes.

The moment something becomes distinguishable,
it becomes vulnerable.

Every organized system now possesses something that can be disrupted.

Something that can be injured.

Something that can fail.

Vulnerability therefore is not an accident of reality.

It is one of the inevitable consequences of organization itself.

Without vulnerability,
there is nothing to preserve.

Without something to preserve,
there is no history.

There is no becoming.

The Architecture therefore recognizes vulnerability as one of reality's great organizational thresholds.

The Scientific Inheritance

Every scientific discipline encounters vulnerability.

Astrophysics teaches that stars continually resist gravitational collapse.

Their organization is never passive.

Hydrostatic equilibrium exists because inward gravitational attraction is continually balanced by outward pressure generated through nuclear fusion.

Should that organization fail,
the star changes.

Its history changes.

Sometimes the universe itself becomes enriched because of that transformation.

Chemistry teaches that molecules continually encounter conditions capable of altering or breaking chemical bonds.

Biology teaches that every living cell experiences continual molecular damage.

DNA accumulates mutations.

Proteins become damaged.

Membranes deteriorate.

Oxidative processes continually challenge biological organization.

Physiology teaches that organisms continually regulate temperature, hydration, metabolism, immune function, and countless other variables in order to remain alive.

Ecology demonstrates that ecosystems experience drought, fire, predation, competition, disease, and climatic change.

History reminds us that civilizations likewise encounter conflict, famine, environmental change, economic instability, and cultural fragmentation.

Across every discipline,

the mechanisms differ.

The constitutional observation quietly repeats.

Every organized system continually encounters forces capable of diminishing its coherence.

The Architecture inherits this observation before asking what organizational principle it reveals.

Entropy and Organized Reality

Modern thermodynamics provides one of the most profound scientific descriptions of vulnerability.

Within isolated systems,

entropy tends to increase.

Organization naturally becomes increasingly dispersed unless energy continually enters the system.

Living organisms are therefore not exceptions to thermodynamic law.

They are extraordinary demonstrations of how open systems maintain organized structure while continually exchanging matter and energy with their surroundings.

The Architecture receives this scientific inheritance with gratitude.

Entropy is not the enemy of reality.

Entropy is one of the conditions within which organized reality unfolds.

The deeper question therefore becomes:

How do organized systems preserve coherence while fully participating within an entropic universe?

That question quietly unites astrophysics,

biology,

medicine,

ecology,

and civilization.

The Architecture of Vulnerability

The Architecture now introduces another constitutional observation.

The greater the organization,
the greater the responsibility required to preserve it.

Simple systems possess relatively few relationships.

Complex systems possess many.

Every new relationship introduces new possibilities.

Every new relationship also introduces new vulnerabilities.

A single-celled organism possesses fewer regulatory challenges than a multicellular organism.

A village possesses fewer institutional responsibilities than a civilization.

A simple nervous system possesses fewer integrative demands than the human brain.

Complexity therefore expands both capability
and vulnerability.

Reality does not become fragile because it becomes organized.

Reality becomes increasingly valuable because organization becomes increasingly worthy of preservation.

The Principle of Coherence Challenge

Every organized system continually encounters influences capable of strengthening or diminishing its coherence.

Some interactions preserve organization.

Others contribute to fragmentation.

Every moment therefore presents what the Architecture calls a Coherence Challenge.

A coherence challenge is any condition requiring an organized system to preserve sufficient organization for continued participation within reality.

The challenge may arise through physics.

Through chemistry.

Through biology.

Through ecology.

Through society.

Or through the interior life of a human being.

The mechanisms differ.

The constitutional question remains.

How shall coherence be preserved?

Human Vulnerability

Human beings often misunderstand vulnerability.

Many assume vulnerability is weakness.

Reality teaches another lesson.

Parents become vulnerable because they love.

Teachers become vulnerable because they invest themselves in future generations.

Physicians become vulnerable because they accept responsibility for lives entrusted to their care.

Scientists become vulnerable because every experiment carries the possibility of discovering that previous understanding was incomplete.

Every meaningful commitment introduces vulnerability because every meaningful commitment has something worth preserving.

Perhaps this is one of reality's quietest lessons.

We often recognize the value of coherence most clearly when coherence begins to diminish.

Health is most deeply appreciated when illness appears.

Trust becomes precious when it is threatened.

Peace becomes visible when conflict arrives.

Life itself often becomes most vivid when its finitude is honestly acknowledged.

The Architecture therefore recognizes vulnerability not as the opposite of strength.

It is one of the deepest expressions of value.

Only what matters can truly be lost.

Only what is vulnerable can truly be protected.

Only what is protected can become an enduring inheritance.

Mathematics of Vulnerability

Mathematics quietly illuminates vulnerability without reducing its richness.

Control theory studies stability under continual disturbance.

Network theory examines resilience when individual connections fail.

Dynamical systems investigate trajectories that remain organized despite continual change.

Information theory recognizes that communication depends upon preserving meaningful organization despite noise.

Each mathematical discipline reveals a similar insight.

Persistence is never passive.

Organization is continually maintained.

The Architecture inherits these mathematical perspectives because they reveal another constitutional principle.

Vulnerability is not disorder.

Vulnerability is the condition that makes adaptive organization necessary.

The Architect's Reflection

For many years I believed that strength meant becoming less vulnerable.

Reality gradually taught another lesson.

Every meaningful commitment introduces vulnerability.

The physician who cares deeply becomes vulnerable.

The parent who loves deeply becomes vulnerable.

The teacher who gives generously becomes vulnerable.

The scientist who pursues truth honestly becomes vulnerable to correction.

Perhaps vulnerability is not the opposite of strength.

Perhaps vulnerability is the price of participating fully in reality.

Only what matters can truly be lost.

Only what is vulnerable can truly be protected.

Only what is protected can become an enduring inheritance.

Quiet Reflection

Consider one relationship that has profoundly shaped your life.

Ask yourself,

Why does it matter?

Because it has never been vulnerable?

Or because it remained faithful despite vulnerability?

Remain with that question.

The deepest forms of meaning rarely emerge where nothing is at risk.

They emerge where something precious continues to be preserved.

Crossing the Next Arch

Every vulnerable system now confronts another necessity.

It must respond.

Some responses fail.

Some preserve organization.

Some strengthen organization.

The Architecture therefore proceeds naturally toward its next constitutional principle.

Adaptation.

Vulnerability alone explains nothing.

The question that follows vulnerability echoes throughout the history of reality.

How does anything endure?

The next chapter begins answering that question.

For the history of organized reality is, in many ways,
the history of adaptation itself.

Cathedral Thread

Constitutional Principle Established

Every organized identity becomes vulnerable because it has become worthy of preservation.

Chapter One established:

Reality precedes explanation.

Chapter Two established:

Understanding deepens through relationship.

Chapter Three established:

Identity emerges through differentiation.

Chapter Four established:

Boundaries preserve identity.

Chapter Five now establishes:

Vulnerability reveals value and calls forth adaptation.

The next chapter asks the inevitable constitutional question.

How does organized reality endure?

The Cathedral now opens its sixth chamber.

Chapter Six

Adaptation

The Discipline of Enduring Change

"To endure is not to resist change. It is to preserve identity while responding wisely to change."

Guiding Question

How do vulnerable systems preserve their identity through continual change, and why is adaptation one of reality's most enduring organizational disciplines?

The previous chapter established that every organized system becomes vulnerable.

Boundaries preserve identity.

Vulnerability reveals value.

Yet vulnerability alone cannot explain persistence.

The unavoidable question now emerges.

How does anything endure?

How does a star remain stable for billions of years?

How does a living cell continually repair itself despite molecular damage?

How does a forest recover after fire?

How does a civilization preserve itself while history continually changes around it?

How does a human life remain faithful to itself while continually growing?

Reality answers this question through one of its most remarkable organizational achievements.

Adaptation.

The history of organized reality is, in many respects, the history of successful adaptation.

Without adaptation,
organization disappears.

With adaptation,
organization becomes history.

The Architecture therefore recognizes adaptation as one of the great constitutional disciplines through which coherent existence persists.

Scientific Foundations

This chapter stands upon the inheritance of empirical science.

Its principal foundations include:

- Evolutionary Biology
- Physiology
- Neuroscience
- Immunology
- Ecology
- Systems Biology
- Engineering
- Control Theory

Across these disciplines,

persistent systems continually respond to changing conditions while preserving recognizable organization.

Evolutionary biology demonstrates adaptation across generations.

Physiology continually regulates internal conditions.

The nervous system reorganizes itself through neural plasticity.

The immune system remembers previous encounters.

Ecosystems recover following disturbance.

Engineering designs adaptive control systems capable of continual correction.

The mechanisms differ profoundly.

The recurring observation remains.

Persistent systems continually adapt.

The Architecture inherits this scientific observation before asking what broader organizational principle it reveals.

Mathematical Foundations

Mathematics provides powerful languages through which adaptation becomes increasingly intelligible.

Among them are:

- Differential Equations
- Feedback Theory
- Control Theory
- Dynamical Systems
- Network Theory
- Information Theory

These mathematical disciplines investigate systems that remain recognizable despite continual change.

None assumes perfect equilibrium.

Instead,

they describe continual adjustment,

continual regulation,

continual correction.

Persistence therefore appears not as stillness,

but as organized change.

The Architecture receives these mathematical languages with gratitude.

Architectural Recognition

Only after empirical science and mathematics have spoken does the Architecture introduce its sixth constitutional recognition.

Enduring systems preserve identity through organized adaptation.

Adaptation should never be confused with random change.

Everything changes.

Adaptation preserves.

The Architecture therefore recognizes adaptation as disciplined participation with reality rather than resistance to it.

Adaptation

The Discipline of Enduring Change

The previous chapter established that every organized system becomes vulnerable.

Boundaries preserve identity.

Yet those same boundaries expose every organized system to continual challenge.

The constitutional question therefore becomes unavoidable.

How does anything endure?

How does a star remain stable for billions of years?

How does a living cell continually repair itself despite continual molecular damage?

How does a forest recover after disturbance?

How does a civilization preserve itself while history continually changes around it?

How does a human life remain faithful to itself while continually growing?

Reality answers this question with one of its most remarkable achievements.

Adaptation.

The history of organized reality is, in many respects,
the history of successful adaptation.

Without adaptation,
organization disappears.

With adaptation,
organization becomes history.

The Architecture therefore recognizes adaptation as one of reality's great constitutional disciplines.

The Scientific Inheritance

Every scientific discipline encounters adaptation.

Evolutionary biology demonstrates that populations gradually change through inherited variation acted upon by natural selection.

Characteristics that improve survival and reproduction become increasingly likely to contribute to future generations.

Adaptation therefore unfolds across immense spans of biological time.

Physiology reveals another remarkable form of adaptation.

The human body continually regulates temperature.

Blood pressure.

Blood chemistry.
Energy balance.
Immune activity.
Hormonal communication.
Repair mechanisms.

These regulatory systems preserve life despite continual environmental variation.

Neuroscience reveals extraordinary adaptability through neural plasticity.

The nervous system continually reorganizes itself in response to learning, experience, injury, and memory.

Immunology demonstrates adaptive immunity.

The immune system remembers previous encounters.

Future responses become faster, more precise, and increasingly effective.

Ecology reveals adaptation across entire ecosystems.

Following disturbance,
communities reorganize.
Species redistribute.
New forms of equilibrium emerge.

Engineering likewise develops adaptive control systems capable of continually responding to changing conditions.

Across every discipline,
the mechanisms differ profoundly.
The recurring observation remains.
Persistent systems continually adapt.

The Architecture inherits that scientific observation before asking what organizational principle it reveals.

Adaptation Is Organized Change

Adaptation should never be confused with change itself.

Everything changes.
Stars change.
Mountains change.
Species change.
Civilizations change.
Change alone explains very little.
Adaptation is different.
Adaptation is organized response.

Change becomes adaptation only when sufficient organization is preserved for continued participation within reality.

This distinction becomes one of the constitutional principles of Coherence Architecture.

Random change fragments.

Organized change preserves.

The Architecture therefore recognizes adaptation as the disciplined preservation of identity through intelligent participation in changing conditions.

Dynamic Stability

Science repeatedly demonstrates that enduring systems are rarely static.

Stars continually convert hydrogen into heavier elements.

Living organisms continually replace damaged molecules.

Forests continually regenerate.

Brains continually reorganize neural networks.

Civilizations continually reinterpret inherited knowledge.

Stability therefore should never be mistaken for immobility.

The deepest forms of stability emerge through continual movement.

Hydrostatic equilibrium illustrates this beautifully.

Gravity continually pulls inward.

Fusion continually pushes outward.

The star persists because neither process disappears.

Its coherence emerges through their continual relationship.

The Architecture recognizes this as one of reality's profound constitutional lessons.

Endurance is organized change.

Mathematics of Adaptation

Mathematics has developed remarkable tools for describing adaptation.

Control theory investigates regulation.

Feedback theory studies continual adjustment.

Dynamical systems describe evolving trajectories.

Network theory examines resilience.

Information theory investigates continual communication.

Each mathematical discipline describes systems that remain recognizable despite continual change.

None assumes perfect equilibrium.

Instead,

they describe continual participation,

continual correction,
continual regulation.

The Architecture receives these mathematical achievements because they illuminate adaptation without reducing it.

Adaptation Across Reality

Adaptation appears throughout organized reality.

Stars continually regulate internal pressure.

Cells continually repair damaged proteins.

Immune systems continually distinguish self from non-self.

Brains continually refine neural pathways.

Families continually adjust to changing generations.

Communities continually reorganize around new circumstances.

Scientific knowledge continually corrects itself through better evidence.

Education continually prepares each generation for realities previous generations could not fully anticipate.

Reality therefore demonstrates adaptation across every level presently known to science.

The mechanisms differ.

The organizational principle remains.

Persistent systems remain teachable.

The Principle of Adaptive Coherence

The Architecture therefore proposes another constitutional principle.

Every organized system possessing boundaries must continually adapt to changing coherence challenges if it is to preserve sufficient organization for continued existence.

Adaptation is not exceptional.

Adaptation is one of reality's enduring organizational disciplines.

The greater the complexity,
the greater the adaptive responsibility.

Adaptation and Human Becoming

Human adaptation extends far beyond biology.

We adapt intellectually through learning.

Emotionally through resilience.

Socially through cooperation.
Scientifically through discovery.
Morally through reflection.
Spiritually through humility.
Civilizations adapt through education.
Knowledge itself becomes one of humanity's greatest adaptive achievements.
Every scientific discovery strengthens future generations.
Every mathematical insight enlarges humanity's descriptive power.
Every faithful teacher extends civilization's adaptive capacity.
Education therefore becomes one of humanity's highest forms of adaptation.
A civilization that ceases learning gradually loses its capacity to respond wisely to reality.

The Architect's Reflection

Many people imagine adaptation means becoming someone different.
Reality suggests something quieter.
Healthy adaptation preserves identity while expanding possibility.
A river changes every moment.
Yet it remains the same river.
A tree continually grows.
Yet it remains recognizably itself.
Human beings mature through experience.
Yet our deepest values often become clearer rather than weaker.
Perhaps genuine growth is not abandoning ourselves.
Perhaps it is becoming more faithful to what reality has always been inviting us to become.

Quiet Reflection

Think about one challenge that permanently changed your life.
At the time,
it may have felt like loss.
Looking back,
did it merely change you?
Or did it teach you something that became part of your character?
Adaptation often appears painful while it is occurring.
Only later do we recognize that it became one of the architects of our lives.
Remain with that thought.

The strongest foundations are often laid beneath the greatest pressures.

Crossing the Next Arch

Adaptation introduces another necessity.

No organized system can respond to conditions it cannot detect.

The quality of adaptation therefore depends upon the quality of perception.

As environments become increasingly dynamic,
organized systems require something new.

They must learn what deserves their attention.

The Architecture therefore proceeds naturally toward its next constitutional principle.

Attention.

For adaptation begins with detection.

And every enduring intelligence first learns what is worthy of noticing.

Cathedral Thread

Constitutional Principle Established

Enduring systems preserve identity through organized adaptation.

Chapter One established:

Reality precedes explanation.

Chapter Two established:

Understanding deepens through relationship.

Chapter Three established:

Identity emerges through differentiation.

Chapter Four established:

Boundaries preserve identity.

Chapter Five established:

Vulnerability reveals value and calls forth adaptation.

Chapter Six now establishes:

Adaptation preserves identity through organized participation with change.

The next chapter asks another constitutional question.

How does adaptation learn what deserves its response?

The Cathedral now opens its seventh chamber.

Chapter Seven

Attention

The First Discipline of Coherence

"Every adaptive system must first learn what deserves its response."

Guiding Question

How do adaptive systems distinguish what matters, and why does attention become one of reality's first great disciplines of organized intelligence?

The previous chapter established that every organized system must continually adapt if it is to preserve its identity.

Yet adaptation introduces another challenge.

No system can respond equally to everything.

Reality continually presents more possibilities than any organized system can fully process.

Every environment contains countless interactions.

Every moment contains innumerable changes.

Every living organism encounters far more information than it can respond to simultaneously.

Every civilization faces more challenges than it can address at once.

Organization therefore requires another remarkable achievement.

Selection.

This chapter explores one of reality's most profound organizational disciplines.

Attention.

Attention is not merely concentration.

Attention is the organized allocation of limited resources toward conditions most relevant to preserving coherence.

Without attention,

adaptation becomes impossible.

The Architecture therefore recognizes attention as one of the first great constitutional disciplines through which organized intelligence begins to emerge.

Scientific Foundations

This chapter stands upon the inheritance of empirical science.

Its principal foundations include:

- Neuroscience
- Cognitive Science
- Evolutionary Biology
- Information Theory
- Systems Biology
- Behavioral Ecology
- Psychology

Across these disciplines,

organized systems continually distinguish relevant information from overwhelming environmental complexity.

The human nervous system filters enormous quantities of sensory input.

Selective attention allocates limited cognitive resources.

Working memory possesses finite capacity.

The immune system distinguishes self from non-self.

Plants orient toward light.

Roots extend toward water and nutrients.

Predators detect movement.

Prey detect danger.

Cells recognize chemical gradients.

DNA repair systems identify damaged genetic sequences requiring correction.

The mechanisms differ profoundly.

The recurring observation remains.

Persistent systems continually distinguish what matters from what does not.

The Architecture inherits this scientific observation before asking what organizational principle it reveals.

Mathematical Foundations

Mathematics likewise provides powerful languages through which attention becomes increasingly intelligible.

Among them are:

- Information Theory
- Signal Detection Theory
- Probability
- Network Theory
- Control Theory
- Decision Theory

Information theory distinguishes meaningful signal from surrounding noise.

Signal detection theory investigates decision-making under uncertainty.

Probability estimates competing possibilities.

Network theory identifies highly connected pathways through which information preferentially flows.

Control theory continually updates system behavior according to incoming information.

Each mathematical language contributes another perspective.

Attention is not random.

Attention organizes information.

The Architecture receives these mathematical languages with gratitude because they illuminate one of reality's most remarkable organizational achievements.

Architectural Recognition

Only after empirical science and mathematics have spoken does the Architecture introduce its seventh constitutional recognition.

Attention is the disciplined selection of information relevant to preserving coherent participation within reality.

Attention is not yet awareness.

Attention is not yet understanding.

Attention prepares both.

The Architecture therefore recognizes attention as the first great discipline through which organized systems begin directing their adaptive participation within reality.

Attention

The First Discipline of Coherence

The previous chapter established that organized systems continually encounter changing conditions.

Every boundary introduces vulnerability.

Every vulnerability requires adaptation.

Yet adaptation introduces another constitutional question.

How does any organized system determine what deserves its response?

Reality continually presents more information than any system can completely process.

Every environment contains innumerable interactions.

Every organism encounters more sensory information than it can fully evaluate.

Every civilization encounters more challenges than it can simultaneously address.

Organization therefore requires another remarkable achievement.

Selection.

Reality begins distinguishing what matters from what does not.

The Architecture calls this organizational discipline

Attention.

Attention is not merely concentration.

Attention is the disciplined allocation of limited organizational resources toward preserving future coherence.

Without attention,

adaptation becomes increasingly impossible.

Without adaptation,

coherence gradually diminishes.

The Architecture therefore recognizes attention as one of reality's first great disciplines of organized intelligence.

The Economy of Attention

Reality quietly teaches another remarkable lesson.

Every organized system possesses limited resources.

Energy is finite.

Time is finite.

Processing capacity is finite.

No organism,

no nervous system,

no civilization,

can respond equally to everything.

Attention therefore becomes one of reality's most valuable organizational resources.

Where attention is wisely invested,

adaptation improves.

Where adaptation improves,

coherence strengthens.

The Architecture therefore recognizes another constitutional principle.

Attention becomes the economy through which organized systems invest limited resources toward preserving future coherence.

This principle quietly appears throughout nature.

Throughout biology.

Throughout civilization.

Throughout human life.

Attention Throughout Biology

The principle extends far beyond the human nervous system.

The immune system continually distinguishes self from non-self.

Plants orient leaves toward available light.

Roots extend toward water and nutrients.

Predators detect movement.

Prey detect danger.

Cells recognize chemical gradients.

DNA repair systems identify damaged genetic sequences requiring correction.

Across biology,

different mechanisms accomplish remarkably similar organizational work.

Persistent systems continually distinguish relevant information from background complexity.

The mechanisms differ profoundly.

The organizational principle remains.

Attention preserves adaptive possibility.

Mathematics and Information

Information theory provides one of humanity's most remarkable mathematical languages for understanding attention.

Communication requires distinguishing meaningful signals from surrounding noise.

Signal detection theory investigates decision-making under uncertainty.

Probability estimates competing possibilities.

Network theory identifies highly connected pathways through which information preferentially flows.

Control theory continually updates system behavior according to incoming information.

Each mathematical language contributes another perspective.

Attention is not random.

Attention organizes information.

The Architecture receives these mathematical achievements because they illuminate one of reality's most universal organizational disciplines.

Meaningful organization depends upon meaningful selection.

The Architecture of Attention

The empirical sciences investigate the mechanisms through which attention operates.

The Architecture asks another question.

What organizational significance does attention introduce into reality?

Reality presents possibilities.

Attention selects.

Selection prepares adaptation.

Adaptation strengthens coherence.

Without attention,
experience remains increasingly chaotic.

With disciplined attention,
experience gradually becomes organized.

The Architecture therefore understands attention not as passive observation, but as active participation.

Every act of attention reorganizes future possibility.

The First Symbolic Recognition

The scientific disciplines investigate the mechanisms through which attention operates.

The Architecture now introduces its first symbolic organizational relationship.

Let:

A represent effective attention.

R represent relevant information.

N represent competing informational noise.

The Architecture symbolically recognizes:

$A \propto R / N$

As relevant information becomes increasingly distinguished from competing informational noise,

adaptive capacity generally increases.

As meaningful selection deteriorates,

organized participation becomes increasingly difficult to preserve.

This symbolic relationship is not proposed as an empirical neuroscientific law.

It is an architectural recognition summarizing an organizational principle repeatedly encountered across multiple scientific disciplines.

The empirical mechanisms remain the work of neuroscience,

cognitive science,

information theory,

and systems biology.

The Architecture recognizes the recurring organizational pattern they collectively reveal.

From Attention to Awareness

Attention alone does not constitute awareness.

A surveillance camera records.

An immune system detects.

A sunflower follows the sun.

Each demonstrates organized selection.

Awareness introduces something fundamentally new.

Experience begins becoming internally represented.

The Architecture therefore recognizes another constitutional progression.

Reality
↓
Boundary
↓
Vulnerability
↓
Adaptation
↓
Attention
↓
Awareness

Attention prepares awareness.

Repeated attention organizes experience.

Organized experience prepares learning.

Learning gradually transforms information into understanding.

Nothing has been skipped.

Everything has been earned.

Attention and Human Life

The same constitutional principle quietly shapes every human life.

We gradually become what we repeatedly attend to.

The scientist attends to evidence.

The physician attends to symptoms.

The teacher attends to understanding.

The engineer attends to structure.

The musician attends to harmony.

The parent attends to a child.

The citizen attends to society.

Civilizations themselves reveal their priorities through collective attention.

Where attention is wisely invested,
future generations inherit stronger foundations.

Where attention becomes continually fragmented,
coherence gradually weakens.

Attention therefore possesses moral significance.

Not because attention itself is moral.

Because attention determines what becomes possible.

The Architect's Reflection

Many people assume wisdom begins with knowledge.

Reality gradually taught another lesson.

Wisdom often begins much earlier.

It begins with learning what deserves attention.

The most remarkable teachers I have known were not simply knowledgeable.

They noticed.

They noticed confusion before questions were asked.

They noticed suffering before words were spoken.

They noticed possibilities others overlooked.

Perhaps attention is one of the earliest quiet expressions of compassion.

Before we help another person,
we must first truly notice them.

Quiet Reflection

Pause.

Think about the people who most shaped your life.

What distinguished them?

Perhaps they possessed extraordinary knowledge.

Perhaps not.

More often,
they gave something rarer.

Their attention.

For a brief moment,
you became more important than everything competing for their awareness.

Remain with that thought.

Attention may be one of the first gifts through which coherent relationships begin.

Crossing the Next Arch

Attention continually selects.

Selection organizes experience.

Experience gradually becomes internally represented.

A new constitutional threshold now appears.

The Architecture must ask another question.

What happens when organized systems not only attend,
but begin experiencing the reality to which they attend?

The next chapter explores one of the most remarkable developments presently known within the history of life.

Awareness.

For the first time,
reality begins becoming aware of itself through the living systems it has produced.

Cathedral Thread

Constitutional Principle Established

Attention is the disciplined allocation of limited organizational resources toward preserving coherent participation within reality.

Chapter One established:

Reality precedes explanation.

Chapter Two established:

Understanding deepens through relationship.

Chapter Three established:

Identity emerges through differentiation.

Chapter Four established:

Boundaries preserve identity.

Chapter Five established:

Vulnerability reveals value.

Chapter Six established:

Adaptation preserves identity through organized change.

Chapter Seven now establishes:

Attention determines where adaptation is invested.

The next chapter asks another constitutional question.

What changes when organized systems begin experiencing the reality to which they attend?

The Cathedral now opens its eighth chamber.

Chapter Eight

Awareness

When Reality Begins to Know Itself

"The universe does not merely become increasingly organized. Through living systems, it becomes capable of awareness."

Guiding Question

How does organized attention become awareness, and why does awareness represent one of reality's greatest organizational thresholds?

The previous chapter established that adaptive systems continually distinguish what deserves their attention.

Attention preserves adaptive possibility.

Yet another constitutional question naturally follows.

What happens when attention becomes sufficiently organized that reality is no longer merely detected,

but experienced?

This chapter explores one of the most remarkable developments presently known within the history of life.

Awareness.

The Architecture approaches this subject with the same discipline that has guided every previous chapter.

First,

we inherit.

Then,

we recognize.

Only afterward,

do we reflect.

Scientific Foundations

This chapter stands upon the inheritance of empirical science.

Its principal foundations include:

- Neuroscience
- Cognitive Science
- Comparative Biology
- Evolutionary Biology
- Psychology
- Information Theory

Modern neuroscience continues investigating one of biology's deepest questions.

How do nervous systems generate awareness?

Remarkable progress has already been made.

Sensory systems continually receive information from the external world.

Neural networks process that information.

Attention selectively amplifies particular signals.

Memory integrates previous experience.

Prediction continually compares expectation with incoming information.

These processes contribute to increasingly sophisticated forms of awareness.

Comparative biology further demonstrates that awareness exists in many degrees.

Single-celled organisms detect environmental gradients.

Plants respond to light, gravity, moisture, and chemical signals.

Animals demonstrate increasingly complex perception, learning, navigation, communication, and social behavior.

Mammals exhibit rich emotional and cognitive capacities.

Human beings presently display the most elaborate forms of reflective awareness known to science.

The mechanisms remain subjects of active scientific investigation.

The Architecture receives this scientific work with gratitude.

It does not attempt to replace it.

Mathematical Foundations

Mathematics likewise provides several complementary languages through which awareness becomes increasingly intelligible.

Among them are:

- Information Theory

- Network Theory
- Complexity Theory
- Predictive Modeling
- Bayesian Inference
- Systems Theory

Information alone does not constitute awareness.
Communication.
Selection.
Prediction.
Integration.
Each contributes another aspect of organized experience.
Neural networks likewise illustrate remarkable mathematical organization.
Billions of neurons communicate through dynamic patterns of connectivity.
No individual neuron explains awareness.
Organization does.

Network theory demonstrates that increasingly rich organization emerges through increasingly rich relationships.

The Architecture receives these mathematical languages because they illuminate the conditions through which awareness becomes increasingly possible.

Architectural Recognition

Only after empirical science and mathematics have spoken does the Architecture introduce its eighth constitutional recognition.

Awareness emerges when organized attention becomes organized experience.

The Architecture does not propose a new neuroscientific theory of consciousness.

Nor does it attempt to explain subjective experience through symbolism alone.

Rather,

it recognizes that empirical science repeatedly demonstrates an increasing integration of attention,

memory,
prediction,
learning,
and organization.

Awareness becomes one of reality's great organizational thresholds.

Awareness

When Reality Begins to Know Itself

The previous chapter established that adaptation requires attention.

Organized systems continually distinguish relevant information from overwhelming environmental complexity.

Selection preserves adaptive possibility.

Yet another constitutional question naturally follows.

What happens when attention becomes sufficiently organized that reality is no longer merely detected,

but experienced?

This question introduces one of the most remarkable developments presently known within the history of life.

Awareness.

The Architecture approaches this subject with the same discipline that has guided every previous chapter.

First,

we inherit.

Then,

we recognize.

Only afterward,

do we reflect.

Evolutionary Deepening

Awareness did not suddenly appear.

Like every previous organizational achievement,

it emerged gradually.

Sensitivity preceded perception.

Perception prepared attention.

Attention prepared awareness.

Awareness expanded through learning.

Learning deepened memory.

Memory strengthened prediction.

Prediction enlarged adaptive intelligence.

Evolution therefore reveals awareness as a continuum rather than an isolated event.

Each new level inherited previous organization while introducing genuinely new capacities.

The Architecture recognizes this gradual deepening as one of reality's most beautiful examples of recursive becoming.

The Architecture of Awareness

The empirical sciences investigate the mechanisms through which awareness emerges.

The Architecture asks another question.

What organizational significance does awareness introduce into reality?

Before awareness,

organized systems respond.

With awareness,

organized systems begin experiencing.

Reality acquires an interior dimension.

Experience begins.

Learning accelerates.

Prediction improves.

Choice expands.

The future begins influencing present behavior.

The Architecture therefore recognizes awareness as one of reality's greatest organizational thresholds.

Not because it suspends biology.

Because it enlarges biology.

The Principle of Reflective Organization

With awareness,

organization no longer remains exclusively reactive.

It becomes increasingly reflective.

Living systems begin monitoring themselves.

Monitoring relationships.

Monitoring environments.

Monitoring future possibilities.

Reflection gradually emerges from repeated awareness.

Reflection permits deliberate learning.

Deliberate learning enlarges adaptation.

Adaptation strengthens coherence.

The Architecture therefore recognizes another constitutional progression.

Reality

↓

Attention

↓

Awareness

↓

Reflection

↓

Learning

↓

Wisdom

Each level inherits those before it while introducing genuinely new organizational possibilities.

Nothing is discarded.

Everything is carried forward.

The First Interior Dimension

One of the remarkable consequences of awareness is that reality begins participating in itself through experience.

The Architecture uses this statement carefully.

It does not suggest that the universe possesses one universal consciousness.

Nor does it propose awareness as an independent physical force.

Rather,

it recognizes an empirical observation.

Through living systems,

reality becomes capable of internally representing aspects of itself.

The world is no longer merely encountered.

It is experienced.

This marks one of the great thresholds in the continuing history of organized existence.

A Symbolic Recognition

The Architecture now introduces another symbolic organizational relationship.

Let

W

represent organized awareness.

A

represent disciplined attention.

L

represent accumulated learning.

M

represent integrated memory.

The Architecture symbolically recognizes:

$W = f(A, L, M)$

This symbolic relationship is not proposed as an empirical neuroscientific law.

It summarizes an organizational recognition repeatedly encountered throughout biology and cognitive science.

Awareness deepens as attention,

learning,

and memory become increasingly integrated.

The mechanisms remain the work of neuroscience.

The Architecture recognizes the recurring organizational pattern they collectively reveal.

Human Awareness

Human awareness introduces extraordinary possibilities.

We become aware of ourselves.

Aware of others.

Aware of history.

Aware of mortality.

Aware of responsibility.

Aware of future generations.

The Architecture does not claim that these capacities appeared suddenly.

They mature gradually.

Education enlarges awareness.

Experience enlarges awareness.

Scientific discovery enlarges awareness.

Compassion enlarges awareness.

Every act of faithful attention gradually transforms how reality is experienced.

Awareness therefore becomes one of humanity's greatest inheritances.

Awareness and Coherence

Attention selected information.

Awareness begins organizing experience.

Experience gradually becomes meaningful.

Meaning strengthens learning.

Learning strengthens future awareness.

A remarkable recursive process begins.

The Architecture recognizes this as one of reality's most elegant organizational cycles.

Awareness continually improves the capacities that produced it.

Learning reorganizes attention.

Attention deepens awareness.

Awareness enlarges learning.

Coherence strengthens.

The cycle continues.

The Architect's Reflection

When I first began thinking about awareness,

I imagined it primarily as knowledge.

Years of listening taught me something quieter.

Awareness often begins with presence.

The most remarkable physicians I have known listened before speaking.

The greatest teachers noticed before explaining.

The wisest parents understood before correcting.

Perhaps awareness is not simply knowing more.

Perhaps awareness begins by seeing more faithfully.

Reality has always been speaking.

Awareness gradually learns how to listen.

Quiet Reflection

Sit quietly.

Notice your breathing.

Notice sounds.

Notice light.

Notice thoughts arising.

Do not immediately judge them.
Simply observe.
Awareness often begins not by adding more experience,
but by becoming fully present to the experience already unfolding.
Remain there for a moment.
Reality has been here all along.

Crossing the Next Arch

Awareness transforms experience.
Yet experience alone does not preserve itself.
For awareness to become civilization,
another remarkable organizational achievement must emerge.
Learning.
The next chapter begins exploring how repeated awareness gradually becomes organized understanding.
For awareness experiences reality.
Learning preserves that experience.
And through learning,
the future begins inheriting the past.

Cathedral Thread

Constitutional Principle Established
Awareness emerges when organized attention becomes organized experience.
Chapter One established:
Reality precedes explanation.
Chapter Two established:
Understanding deepens through relationship.
Chapter Three established:
Identity emerges through differentiation.
Chapter Four established:
Boundaries preserve identity.
Chapter Five established:
Vulnerability reveals value.
Chapter Six established:
Adaptation preserves identity through organized change.
Chapter Seven established:
Attention directs adaptive participation.

Chapter Eight now establishes:
Awareness transforms organized attention into lived experience.
The next chapter asks another constitutional question.
How does lived experience become enduring learning?
The Cathedral now opens its ninth chamber.

Chapter Nine

Learning

When Experience Becomes Organization

"Reality continually teaches. Learning is the gradual organization of that teaching into enduring understanding."

Guiding Question

How does awareness become learning, and why does learning represent one of reality's greatest organizational achievements?

The previous chapter established that awareness allows organized systems to experience reality.

Experience alone, however, is not sufficient.

Experience may be forgotten.

Experience may be misunderstood.

Experience may remain isolated.

For experience to contribute to future coherence,
something remarkable must occur.

Experience must become learning.

This chapter explores one of reality's greatest organizational achievements.

Learning.

Learning allows the past to influence the future.

It allows successful adaptation to become increasingly likely.

It allows every generation to begin where previous generations ended rather than where they themselves began.

Learning therefore transforms awareness into inheritance.

The Architecture recognizes learning as one of the principal bridges between biology and civilization.

Scientific Foundations

This chapter stands upon the inheritance of empirical science.

Its principal foundations include:

- Neuroscience
- Cognitive Science
- Educational Psychology
- Evolutionary Biology
- Immunology
- Information Theory
- Artificial Intelligence (Learning Systems)

Modern neuroscience demonstrates that learning continually reshapes the brain.

Neural pathways strengthen through repeated activity.

New synaptic connections form.

Existing connections weaken or disappear.

The nervous system continually reorganizes itself according to experience.

Learning therefore is not merely the accumulation of information.

Learning physically reorganizes biological systems.

Cognitive science further reveals that learning continually updates internal models of reality.

Prediction improves.

Error decreases.

Behavior becomes increasingly adaptive.

Evolutionary biology demonstrates another remarkable truth.

Learning extends beyond individuals.

Populations inherit successful adaptations across generations.

Species therefore possess biological histories shaped by countless previous interactions with reality.

Education extends this process still further.

Knowledge acquired by one generation becomes available to the next.

Civilization itself becomes an extraordinary extension of learning across time.

The Architecture receives this scientific inheritance with gratitude.

Mathematical Foundations

Mathematics likewise provides powerful languages through which learning becomes increasingly intelligible.

Among them are:

- Information Theory
- Bayesian Inference
- Probability Theory
- Network Theory
- Recursive Mathematics
- Learning Algorithms

Information theory investigates how uncertainty decreases through acquired information.

Probability continually updates expectation according to new evidence.

Bayesian reasoning illustrates how prior understanding may be refined through subsequent observation.

Network theory demonstrates how repeated connections strengthen organized pathways.

Recursive mathematics illustrates how successive iterations generate increasingly sophisticated organization.

Each mathematical language contributes another perspective.

Together they reveal one remarkable constitutional principle.

Learning organizes the future through the faithful inheritance of the past.

The Architecture receives these mathematical achievements with gratitude.

Architectural Recognition

Only after empirical science and mathematics have spoken does the Architecture introduce its ninth constitutional recognition.

Learning is organized experience that improves future participation within reality.

Experience alone disappears.

Learning preserves.

Learning reorganizes.

Learning prepares.

The Architecture therefore recognizes learning as one of reality's greatest recursive achievements.

Learning

When Experience Becomes Organization

The previous chapter established that awareness allows organized systems to experience reality.

Experience alone, however,

is not sufficient.

Experience may be forgotten.

Experience may remain isolated.

Experience may never influence future behavior.

For experience to contribute to future coherence,

it must become organized.

It must become learning.

Learning therefore represents one of reality's greatest organizational achievements.

It allows the past to influence the future.

It allows successful adaptation to become increasingly likely.

It allows every generation to begin where previous generations ended rather than where they themselves began.

Learning transforms awareness into inheritance.

The Architecture therefore recognizes learning as one of the principal bridges between biology and civilization.

Learning Throughout Reality

The Architecture recognizes learning across many domains.

The immune system remembers previous pathogens.

Future responses become increasingly effective.

Plants alter growth according to environmental conditions.

Animals modify behavior through experience.

Brains continually reorganize.

Scientific communities revise theories according to evidence.

Engineering systems increasingly incorporate adaptive learning algorithms.

Human cultures preserve discoveries through language, writing, education, and institutions.

The mechanisms differ profoundly.

The constitutional observation remains.

Reality continually teaches.

Organized systems increasingly retain what reality has taught.

Mathematics and Learning

Mathematics provides another remarkable language through which learning becomes increasingly intelligible.

Information theory investigates how uncertainty decreases through acquired information.

Probability continually updates expectation according to new evidence.

Bayesian reasoning illustrates how prior understanding may be refined through subsequent observation.

Network theory demonstrates how repeated connections strengthen organized pathways.

Recursive mathematics reveals how successive iterations generate increasingly sophisticated organization.

Each mathematical language contributes another perspective.

Together they illuminate one remarkable constitutional observation.

Learning organizes the future through the faithful inheritance of the past.

The Architecture receives these mathematical languages with gratitude because they reveal another enduring principle.

Learning is not merely accumulation.

Learning is organized transformation.

The Architecture of Learning

The empirical sciences investigate the mechanisms through which learning occurs.

The Architecture asks another question.

What organizational transformation occurs when experience becomes learning?

Experience is temporary.

Learning preserves.

Experience encounters reality.

Learning reorganizes future participation within reality.

The Architecture therefore recognizes learning as organized experience capable of influencing future coherence.

Reality teaches.

Learning remembers.
Every successful lesson becomes part of future organization.
Experience no longer disappears.
It begins participating in what comes next.

Recursive Learning

Learning possesses another remarkable characteristic.
It continually improves itself.
Every lesson changes the learner.
The changed learner now encounters reality differently.
New observations become possible.
New questions emerge.
Better questions generate deeper understanding.
Learning therefore becomes recursive.
Each cycle enlarges the next.
Reality
↓
Experience
↓
Learning
↓
Improved Awareness
↓
Richer Experience
↓
Deeper Learning
The cycle continues throughout life.

The Architecture recognizes this as one of reality's most elegant recursive structures.

A Symbolic Recognition

The Architecture now introduces another symbolic organizational relationship.
Let
L
represent organized learning.
E
represent accumulated experience.

A

represent disciplined attention.

R

represent reflective integration.

The Architecture symbolically recognizes:

$L = f(E, A, R)$

Learning deepens as experience becomes increasingly attended,

reflected upon,

and integrated.

This symbolic relationship is not proposed as an empirical neuroscientific law.

It summarizes an organizational recognition repeatedly encountered throughout neuroscience,

cognitive science,

educational psychology,

and systems theory.

The empirical sciences investigate the mechanisms.

The Architecture recognizes the recurring organizational relationship they collectively reveal.

Learning and Civilization

Perhaps humanity's greatest adaptive achievement is not intelligence alone.

It is education.

Education allows discoveries to survive individual lives.

Every scientific paper.

Every mathematical theorem.

Every physician's experience.

Every engineer's design.

Every teacher's wisdom.

Every parent's guidance.

Every civilization's memory.

These become organized inheritances.

Without education,

every generation would begin almost entirely anew.

Learning therefore extends individual adaptation into civilizational adaptation.

Knowledge accumulates.

Understanding matures.

Humanity gradually becomes capable of asking better questions.

The Principle of Educational Coherence

The Architecture therefore proposes another constitutional principle.

Learning strengthens coherence when reality is inherited more faithfully than it was previously understood.

The purpose of education is therefore not merely information transfer.

Its purpose is coherent inheritance.

Education preserves reality's lessons across generations.

Civilization depends upon that continuity.

The Architect's Reflection

For many years I believed that learning meant acquiring more knowledge.

Reality gradually taught another lesson.

Learning often begins with unlearning.

Every scientific revolution required humanity to release explanations that no longer matched observation.

Every mature teacher eventually discovers that certainty often prevents deeper understanding.

Reality continues teaching those who remain teachable.

Perhaps the greatest learner is not the one who knows the most.

Perhaps the greatest learner is the one who remains most faithful to reality.

Quiet Reflection

Think of one lesson that permanently changed your life.

Not simply information.

A lesson.

Ask yourself,

What reality taught it?

How did that lesson reorganize your future?

Remain with those questions.

Reality has been your teacher far longer than you have recognized.

Every meaningful lesson has become another stone in the architecture of your own life.

Crossing the Next Arch

Learning preserves experience.

Yet learning alone does not guarantee permanence.

What learning acquires,
something else must preserve.

The next chapter explores another extraordinary organizational achievement.

Memory.

For learning prepares understanding.

Memory preserves understanding.

And through memory,
the future begins inheriting the past.

Cathedral Thread

Constitutional Principle Established

Learning is organized experience that improves future participation within reality.

Chapter One established:
Reality precedes explanation.

Chapter Two established:
Understanding deepens through relationship.

Chapter Three established:
Identity emerges through differentiation.

Chapter Four established:
Boundaries preserve identity.

Chapter Five established:
Vulnerability reveals value.

Chapter Six established:
Adaptation preserves identity through organized change.

Chapter Seven established:
Attention directs adaptive participation.

Chapter Eight established:
Awareness transforms attention into lived experience.

Chapter Nine now establishes:
Learning reorganizes the learner.

The next chapter asks another constitutional question.

How is learning preserved beyond the moment in which it occurs?

The Cathedral now opens its tenth chamber.

Chapter Ten

Memory

When Reality Begins to Remember

"Learning transforms experience. Memory transforms learning into inheritance."

Guiding Question

How does learning become enduring memory, and why does memory represent one of reality's greatest organizational achievements?

The previous chapter established that learning reorganizes future participation within reality.

Learning, however,
remains incomplete unless it can be preserved.
Without preservation,
every lesson disappears.
Every adaptation is lost.
Every discovery vanishes.
Every generation begins almost entirely anew.

Reality therefore demonstrates another extraordinary organizational achievement.

Memory.

Memory preserves organization through time.

It allows the past to continue participating in the future.

This chapter explores one of the deepest transitions in organized existence.

Memory transforms learning into inheritance.

The Architecture therefore recognizes memory as one of the principal foundations upon which persistence,
education,
history,
and civilization become possible.

Scientific Foundations

This chapter stands upon the inheritance of empirical science.

Its principal foundations include:

- Genetics
- Molecular Biology
- Neuroscience
- Immunology
- Developmental Biology
- Cognitive Science
- Evolutionary Biology

Throughout biology,
memory appears in many remarkable forms.

DNA preserves biological information across generations.

The genetic code transmits instructions accumulated across immense evolutionary timescales.

Cells continually preserve regulatory information through coordinated molecular processes.

The immune system retains memory of previous pathogens, allowing faster and more effective future responses.

Neuroscience demonstrates that memory continually reshapes neural organization.

Experience strengthens some connections.

Others weaken.

Still others disappear.

Learning therefore becomes physically embedded within the nervous system.

Psychology further distinguishes multiple forms of memory.

Sensory memory.

Working memory.

Procedural memory.

Semantic memory.

Episodic memory.

Each contributes differently to adaptive participation.

The mechanisms remain subjects of active scientific investigation.
The Architecture receives these discoveries with gratitude.
They reveal that memory is not merely storage.
Memory is organized preservation.

Mathematical Foundations

Mathematics likewise provides powerful languages through which memory becomes increasingly intelligible.

Among them are:

- Information Theory
- Coding Theory
- Network Theory
- Graph Theory
- Recursive Mathematics
- Dynamical Systems

Information theory investigates reliable preservation of information.
Coding theory studies faithful transmission despite noise.
Network theory reveals persistent organizational pathways.
Recursive mathematics illustrates how successive organization inherits previous organization.
Graph theory demonstrates how increasingly connected knowledge becomes increasingly resilient.
Each mathematical language contributes another perspective.
Together they illuminate another constitutional observation.
Memory is organized continuity through time.
The Architecture inherits these mathematical achievements with gratitude.

Architectural Recognition

Only after empirical science and mathematics have spoken does the Architecture introduce its tenth constitutional recognition.

Memory is organized preservation that allows the past to participate in the future.

Learning acquires.
Memory preserves.
Inheritance becomes possible.
Civilization becomes possible.
Reality therefore begins carrying itself forward through organized remembrance.
The Architecture recognizes memory as one of the great accelerators of becoming.

Memory

When Reality Begins to Remember

The previous chapter established that learning reorganizes experience into increasingly adaptive participation.

Learning, however,
remains incomplete unless it can be preserved.
Without preservation,
every lesson disappears.
Every adaptation is lost.
Every discovery vanishes.

Reality therefore demonstrates another remarkable organizational achievement.

Memory.

Memory preserves organization through time.

The Architecture therefore recognizes memory as one of the principal foundations upon which persistence,
inheritance,
and civilization become possible.

Memory allows the past to continue participating in the future.
Without memory,
history becomes impossible.
Without history,
there can be no civilization.

Memory Throughout Reality

The Architecture recognizes memory across many domains of organized reality.

Geological formations preserve Earth's history.

Tree rings preserve environmental history.

Ice cores preserve climatic history.

Fossils preserve biological history.

Languages preserve cultural memory.

Libraries preserve intellectual memory.

Scientific literature preserves empirical memory.

Mathematics preserves logical memory.

Civilizations themselves become living repositories of accumulated understanding.

The mechanisms differ profoundly.

The constitutional observation remains.

Reality continually develops increasingly sophisticated ways of preserving organized information through time.

Mathematics and Memory

Mathematics provides several powerful languages through which memory becomes increasingly intelligible.

Information theory investigates information preserved through communication.

Coding theory investigates reliable transmission despite noise.

Network theory studies persistent patterns of connectivity.

Recursive mathematics illustrates how successive organization inherits previous organization.

Graph theory demonstrates how increasingly connected knowledge becomes increasingly resilient.

Each mathematical language contributes another perspective.

Memory is not merely stored information.

Memory is organized continuity.

The Architecture receives these mathematical achievements with gratitude because they illuminate one of reality's most enduring organizational disciplines.

The Architecture of Memory

The empirical sciences investigate how memory is formed.

The Architecture asks another question.

What organizational significance does memory introduce into reality?

Memory preserves successful organization.

Memory reduces unnecessary repetition.

Memory strengthens future adaptation.

Memory therefore enlarges coherence.

Every organized system capable of remembering possesses possibilities unavailable to systems continually beginning anew.

Reality itself appears to move increasingly toward organized inheritance.

The Architecture recognizes memory as one of the great accelerators of becoming.

Recursive Inheritance

Memory introduces another remarkable organizational principle.

Inheritance.

Every new generation begins with more than immediate experience.

Genes are inherited.

Languages are inherited.

Cultures are inherited.

Mathematics is inherited.

Science is inherited.

Wisdom is inherited.

Even the questions humanity asks are often inherited from previous generations.

Reality therefore progresses through recursive inheritance.

The future continually builds upon organized memory.

Inheritance does not require repetition.

Inheritance continually generates novelty.

Every generation preserves.

Every generation transforms.

Every generation contributes.

A Symbolic Recognition

The Architecture now introduces another symbolic organizational relationship.

Let

M

represent organized memory.

L

represent accumulated learning.

I

represent inherited organization.

P

represent preserved patterns.

The Architecture symbolically recognizes:

$M = f(L, I, P)$

Memory strengthens as learning becomes increasingly organized,

preserved,

and transmitted.

This symbolic relationship is not proposed as an empirical biological law.

It summarizes an organizational recognition repeatedly encountered across genetics,

neuroscience,

immunology,

education,

and information theory.

The empirical sciences investigate the mechanisms.

The Architecture recognizes the recurring organizational relationship they collectively reveal.

Memory and Education

Perhaps the highest expression of organized memory is education.

Education allows discoveries to survive individual lives.

The scientist teaches the student.

The physician teaches the resident.

The engineer teaches the apprentice.

The parent teaches the child.

The teacher preserves civilization through organized memory.

Education therefore becomes humanity's greatest memory system.

Without education,

knowledge continually disappears.

With education,

understanding accumulates.

Civilization itself becomes recursive memory extending across generations.

The Architecture therefore recognizes education as one of reality's highest organizational achievements.

The Principle of Coherent Inheritance

The Architecture therefore proposes another constitutional principle.

Memory strengthens coherence when faithfully preserved understanding becomes available to future organization.

The purpose of memory is not merely preservation.

Its purpose is preparation.

Memory prepares adaptation.

Memory prepares wisdom.

Memory prepares civilization.

Reality therefore continually remembers so that organized becoming may continue.

The Cathedral Equation Begins to Emerge

Throughout the previous chapters,

one quiet pattern has repeatedly appeared.

Reality organizes.

Boundaries preserve.

Adaptation responds.

Attention selects.

Awareness experiences.

Learning transforms.

Memory preserves.

Across every scientific discipline,

organized systems continually preserve themselves against continual forces of fragmentation.

The Architecture recognizes that these recurring observations gradually prepare another insight.

Coherence.

The reader has not yet encountered the Cathedral Equation.

Yet every chapter has quietly prepared it.

The equation will not introduce a new idea.

It will summarize one the reader has already recognized.

That is the purpose of architecture.

Not to invent reality.

To reveal its organization.

The Architect's Reflection

When I first thought about memory,
I imagined recollection.
Reality gradually taught me something deeper.
Memory is civilization's promise to the future.
Every discovery remembered.
Every lesson preserved.
Every act of compassion passed forward.
Every generation inherits far more than information.
It inherits possibility.
Perhaps memory is not simply about the past.
Perhaps memory is reality's way of preparing the future.

Quiet Reflection

Think about someone whose influence continues within your life.
Perhaps they are no longer living.
Yet something they taught remains.
A sentence.
A habit.
A value.
A kindness.
That memory continues shaping decisions long after the moment itself has passed.
Remain with that thought.
The greatest memories do not merely preserve yesterday.
They quietly prepare tomorrow.

Crossing the Next Arch

Memory preserves.
Yet preservation alone does not produce understanding.
Information may be remembered without becoming knowledge.
Facts may accumulate without becoming wisdom.
The next chapter therefore enters another remarkable threshold.
Knowledge.
For memory preserves reality's lessons.

Knowledge organizes those lessons into coherent understanding.
Through knowledge,
humanity begins reading reality more faithfully than ever before.

Cathedral Thread

Constitutional Principle Established

Memory is organized preservation that allows the past to participate in the future.

Chapter One established:
Reality precedes explanation.
Chapter Two established:
Understanding deepens through relationship.
Chapter Three established:
Identity emerges through differentiation.
Chapter Four established:
Boundaries preserve identity.
Chapter Five established:
Vulnerability reveals value.
Chapter Six established:
Adaptation preserves identity through organized change.
Chapter Seven established:
Attention directs adaptive participation.
Chapter Eight established:
Awareness transforms attention into lived experience.
Chapter Nine established:
Learning reorganizes the learner.
Chapter Ten now establishes:
Memory preserves organized learning so the future may inherit it.
The next chapter asks another constitutional question.
How does preserved memory become organized knowledge?
The Cathedral now opens its eleventh chamber.

Chapter Eleven

Knowledge

When Memory Becomes Understanding

"Memory preserves the past. Knowledge organizes the past so that the future may understand reality more faithfully."

Guiding Question

How does preserved memory become organized knowledge, and why does knowledge represent one of humanity's greatest organizational achievements?

The previous chapter established that memory preserves learning.

Experience became learning.

Learning became memory.

Yet memory alone does not necessarily produce understanding.

A library preserves memory.

An archive preserves memory.

A computer stores information.

None alone possesses knowledge.

Something further must occur.

Patterns must become visible.

Relationships must become intelligible.

Understanding must emerge.

This chapter explores one of reality's greatest organizational achievements.

Knowledge.

Knowledge is not merely accumulated information.

Knowledge is organized understanding grounded in reality.

The Architecture therefore recognizes knowledge as one of the principal bridges between memory and wisdom.

Scientific Foundations

This chapter stands upon the inheritance of empirical science.

Its principal foundations include:

- Astronomy
- Physics
- Chemistry
- Biology
- Medicine
- Neuroscience
- Philosophy of Science

Every scientific discipline demonstrates the remarkable transformation of memory into organized understanding.

Astronomers do not merely collect observations.

They organize observations into increasingly coherent models of the universe.

Physicists organize measurements into mathematical laws describing matter,

energy,

motion,

and spacetime.

Chemists organize atomic behavior into periodic relationships,

molecular structures,

and reaction networks.

Biologists organize countless observations into theories of evolution,

genetics,

physiology,

ecology,

and cellular organization.

Medicine continually organizes biological knowledge into increasingly effective approaches to preserving health.

Science therefore demonstrates a profound lesson.

Knowledge emerges when observations become organized into reliable explanations that increasingly correspond to reality.

The Architecture receives this scientific inheritance with gratitude.

Mathematical Foundations

Mathematics provides one of humanity's greatest instruments for organizing knowledge.

Among its principal languages are:

- Geometry
- Algebra
- Calculus
- Probability
- Linear Algebra
- Graph Theory
- Information Theory
- Topology

Geometry organizes space.
Algebra organizes relationships.
Calculus organizes change.
Probability organizes uncertainty.
Graph theory organizes networks.
Information theory organizes communication.
Topology organizes continuity.
None creates reality.
Each faithfully reveals relationships already present within reality.
The Architecture therefore receives mathematics not merely as calculation, but as one of humanity's greatest instruments for organizing understanding.

Architectural Recognition

Only after empirical science and mathematics have spoken does the Architecture introduce its eleventh constitutional recognition.

Knowledge is organized memory that increasingly reveals reality's recurring relationships.

Memory preserves.

Knowledge relates.

Knowledge recognizes.

Knowledge therefore transforms accumulated inheritance into organized understanding.

The Architecture recognizes knowledge as one of reality's great recursive achievements.

Knowledge

When Memory Becomes Understanding

The previous chapter established that memory preserves learning.

Learning, however,
does not by itself produce understanding.

Memory stores.

Knowledge organizes.

A library possesses memory.

An archive preserves history.

A computer stores information.

None alone possesses knowledge.

For knowledge to emerge,
relationships must become visible.

Patterns must become intelligible.

Understanding must arise.

Knowledge therefore represents one of reality's greatest organizational achievements.

It allows humanity not merely to remember reality,
but increasingly to understand it.

Knowledge Across Reality

Knowledge appears wherever organized systems increasingly represent reality.

The immune system acquires knowledge of pathogens.

Brains acquire knowledge through experience.

Scientific communities accumulate knowledge through disciplined investigation.

Cultures preserve knowledge through education.

Civilizations organize knowledge through libraries,
universities,
laboratories,
and institutions.

The mechanisms differ profoundly.

The constitutional observation remains.

Reality teaches.

Organization preserves.

Knowledge expands.

The Architecture therefore recognizes knowledge as one of reality's greatest recursive inheritances.

Every generation begins with knowledge previous generations patiently organized.

Mathematics and Knowledge

Mathematics provides another extraordinary language through which knowledge becomes increasingly intelligible.

Geometry organizes space.

Algebra organizes relationships.

Calculus organizes change.

Probability organizes uncertainty.

Graph theory organizes networks.

Information theory organizes communication.

Topology organizes continuity.

Linear algebra organizes transformations.

Each mathematical discipline contributes a different way of organizing reality.

None creates reality.

Each faithfully reveals relationships already present within it.

The Architecture therefore recognizes mathematics as one of humanity's greatest instruments for transforming preserved memory into organized knowledge.

Without mathematics,
many of reality's deepest relationships would remain hidden.

The Architecture of Knowledge

The empirical sciences investigate the processes through which knowledge develops.

The Architecture asks another question.

What organizational transformation occurs when memory becomes knowledge?

Memory preserves.

Knowledge relates.

Knowledge recognizes patterns.

Knowledge identifies recurring principles.

Knowledge therefore compresses enormous complexity into increasingly coherent understanding.

One equation may summarize thousands of observations.

One biological principle may organize millions of individual events.

One mathematical relationship may illuminate countless physical systems.

Knowledge therefore becomes organized memory capable of guiding future discovery.

The Architecture recognizes knowledge as one of reality's greatest organizational achievements because it continually enlarges humanity's capacity to understand rather than merely remember.

Recursive Organization

Knowledge possesses another remarkable characteristic.

It continually reorganizes itself.

Every discovery changes previous understanding.

Every improved explanation reorganizes earlier explanations.

Knowledge therefore grows recursively.

Observation

↓

Learning

↓

Memory

↓

Knowledge

↓

Improved Observation

↓

Deeper Knowledge

Reality continually teaches.

Knowledge continually reorganizes itself in response.

The Architecture recognizes this recursive refinement as one of civilization's greatest strengths.

Truth welcomes correction.

Only error fears it.

A Symbolic Recognition

The Architecture now introduces another symbolic organizational relationship.

Let

K

represent organized knowledge.

M

represent preserved memory.

R

represent recognized relationships.

I

represent integrated understanding.

The Architecture symbolically recognizes:

$K = f(M, R, I)$

Knowledge deepens as preserved memory becomes increasingly organized through faithful recognition of relationships and their coherent integration.

This symbolic relationship is not proposed as an empirical scientific law.

It summarizes an organizational recognition repeatedly encountered across the empirical sciences.

Astronomy reveals relationships.

Physics reveals relationships.

Chemistry reveals relationships.

Biology reveals relationships.

Mathematics faithfully describes those relationships.

The Architecture recognizes the recurring organizational pattern they collectively reveal.

This is the constitutional discipline of the Cathedral.

First,

empirical discovery.

Second,

mathematical description.

Third,

architectural recognition.

Knowledge and Scientific Integrity

Knowledge does not eliminate uncertainty.

Scientific knowledge remains continually open to refinement.

Every generation inherits improved instruments.
Improved measurements.
Improved mathematics.
Improved evidence.
The Architecture therefore distinguishes between certainty and confidence.
Scientific confidence grows through repeated confirmation.
Scientific certainty remains provisional before reality.
This distinction protects intellectual humility.
Reality remains greater than every explanation humanity constructs.
The greatest strength of science has never been certainty.
Its greatest strength has been its willingness to continue learning.

Knowledge and Civilization

Civilizations rise through organized knowledge.
Agriculture.
Engineering.
Medicine.
Astronomy.
Navigation.
Architecture.
Education.
Justice.
Each depends upon knowledge faithfully preserved across generations.
Civilizations decline when organized knowledge becomes fragmented,
ignored,
or forgotten.

The Architecture therefore recognizes knowledge as one of civilization's primary responsibilities.

Knowledge is never merely inherited.
Knowledge must continually be renewed,
organized,
tested,
and faithfully transmitted.

The Principle of Coherent Knowledge

The Architecture therefore proposes another constitutional principle.

Knowledge strengthens coherence when reality is organized more faithfully than it was previously understood.

The purpose of knowledge is not accumulation.

The purpose of knowledge is orientation.

Knowledge helps humanity read reality more faithfully.

Every generation receives reality.

Knowledge gradually teaches humanity how to read it more clearly.

Toward the Cathedral Equation

Throughout this journey,

the reader has quietly encountered the same recurring observation.

Reality organizes.

Boundaries preserve.

Adaptation responds.

Attention selects.

Awareness experiences.

Learning reorganizes.

Memory preserves.

Knowledge organizes.

Across every scientific discipline,

organized systems continually preserve coherent relationships despite continual change.

The Architecture has not invented this pattern.

The reader has already recognized it.

Soon,

that recognition will receive symbolic expression.

The Cathedral Equation will not introduce a new truth.

It will simply give mathematical voice to one the reader already knows.

The Architect's Reflection

When I first began searching for knowledge,

I believed I was searching for answers.

Years later,

I realized that knowledge often replaces answers with better questions.

Reality continually expands faster than explanation.

Every discovery opens another horizon.

Perhaps knowledge is not measured by the number of conclusions we possess.

Perhaps knowledge is measured by the quality of the questions we have learned to ask.

Quiet Reflection

Consider one belief you once held with complete confidence.

What changed it?

More information?

Or better understanding?

Knowledge rarely grows because facts increase alone.

Knowledge grows because relationships become clearer.

Remain with that thought.

Understanding is often less about collecting more,
and more about seeing reality more faithfully.

Crossing the Next Arch

Knowledge organizes reality.

Yet knowledge alone cannot tell us why anything matters.

Facts accumulate.

Understanding deepens.

Another constitutional question now emerges.

How shall organized knowledge orient human life?

The next chapter enters one of humanity's oldest questions.

Meaning.

For knowledge explains.

Meaning directs.

Together they prepare the Architecture for its final movement toward Coherence,

Wisdom,

Compassion,

and Civilization.

Cathedral Thread

Constitutional Principle Established

Knowledge is organized memory that increasingly reveals reality's recurring relationships.

Chapter One established:

Reality precedes explanation.

Chapter Two established:

Understanding deepens through relationship.

Chapter Three established:

Identity emerges through differentiation.

Chapter Four established:

Boundaries preserve identity.

Chapter Five established:

Vulnerability reveals value.

Chapter Six established:

Adaptation preserves identity through organized change.

Chapter Seven established:

Attention directs adaptive participation.

Chapter Eight established:

Awareness transforms attention into lived experience.

Chapter Nine established:

Learning reorganizes the learner.

Chapter Ten established:

Memory preserves organized learning for the future.

Chapter Eleven now establishes:

Knowledge organizes preserved memory into increasingly faithful understanding of reality.

The next chapter asks another constitutional question.

How does understanding become meaningful participation within reality?

The Cathedral now opens its twelfth chamber.

Chapter Twelve

Meaning

When Understanding Begins to Guide Life

"Science teaches us how reality unfolds. Meaning asks how we shall live within that unfolding."

Guiding Question

How does organized knowledge become meaningful participation within reality, and why does meaning represent one of humanity's greatest organizational thresholds?

The previous chapter established that knowledge organizes preserved memory into increasingly faithful understanding.

Knowledge, however,

does not by itself determine how human beings should live.

Knowledge explains.

Meaning orients.

Knowledge may reveal the age of a star.

Meaning asks why wonder matters.

Knowledge may explain cellular repair.

Meaning asks why life deserves preservation.

Knowledge may illuminate the history of civilizations.

Meaning asks what responsibilities that history places upon the present generation.

Reality therefore presents humanity with two complementary questions.

The first asks,

How does reality work?

The second asks,

How shall we live within reality?

This chapter explores one of humanity's oldest and deepest organizational achievements.

Meaning.

The Architecture recognizes meaning not as an alternative to scientific un-

derstanding,
but as one of its mature consequences.

Scientific Foundations

This chapter stands upon the inheritance of empirical science and the human sciences.

Its principal foundations include:

- Psychology
- Neuroscience
- Anthropology
- Cognitive Science
- History
- Ethics
- Philosophy of Science

The scientific disciplines investigate reality through observation,
experimentation,
mathematical description,
and continual refinement.
Psychology investigates human motivation,
emotion,
purpose,
and meaning-making.
Neuroscience studies the biological processes associated with valuation,
memory,
decision-making,
and social behavior.
Anthropology investigates how cultures construct shared systems of significance.
History reveals how meaning continually shapes civilizations.
The humanities preserve humanity's long conversation concerning beauty,
justice,
responsibility,
hope,

suffering,
and purpose.
These disciplines differ in method.
Each contributes something indispensable.
The Architecture receives every one with gratitude.
Science investigates what reality is.
Human experience continually asks what reality asks of us.

Mathematical Foundations

Mathematics remains essential.

Not because mathematics calculates meaning.

But because mathematics faithfully organizes the measurable relationships within which meaning is lived.

Among its enduring contributions are:

- Information Theory
- Decision Theory
- Network Theory
- Probability
- Systems Theory

These mathematical languages describe communication,
choice,
organization,
uncertainty,
and interaction.
They do not determine meaning.

They prepare the conditions within which meaningful participation becomes possible.

The Architecture therefore receives mathematics exactly where it belongs.
As the language of measurable relationships.
Meaning begins only after mathematics has honestly completed its work.

Architectural Recognition

Only after empirical science,

the human sciences,

and mathematics have spoken does the Architecture introduce its twelfth constitutional recognition.

Meaning is organized understanding that faithfully orients participation within reality.

Knowledge explains.

Meaning directs.

Knowledge reveals.

Meaning responds.

The Architecture therefore recognizes meaning as one of humanity's greatest organizational thresholds.

Meaning

When Understanding Begins to Guide Life

The previous chapter established that knowledge organizes memory into increasingly faithful understanding.

Knowledge, however,

does not by itself determine how human beings should live.

Knowledge explains.

Meaning orients.

Knowledge may reveal the age of the universe.

Meaning asks why wonder matters.

Knowledge may explain biological evolution.

Meaning asks what responsibilities accompany life.

Knowledge may illuminate the history of civilizations.

Meaning asks how future civilizations ought to be built.

Reality therefore presents humanity with two complementary questions.

How does reality work?

How shall we now live within reality?

Neither question diminishes the other.

Together they enlarge human understanding.

Meaning Emerges Through Relationship

Meaning rarely exists in isolation.

Meaning emerges through participation.

Parents discover meaning through children.

Teachers discover meaning through students.

Scientists discover meaning through discovery.

Physicians discover meaning through healing.

Communities discover meaning through service.

Civilizations discover meaning through shared purpose.

The Architecture therefore recognizes that meaning is not detached from reality.

Meaning emerges through the quality of our participation within reality.

The more faithfully we understand reality,

the more responsibly we may participate within it.

Mathematics and Meaning

Mathematics does not calculate meaning.

Nor should it.

Its purpose is different.

Mathematics faithfully describes measurable relationships.

Meaning interprets how human beings participate within those relationships.

This distinction remains essential.

Geometry reveals structure.

Information theory reveals communication.

Network theory reveals organization.

Probability reveals uncertainty.

Systems theory reveals interaction.

The Architecture receives these mathematical achievements exactly where they belong.

They faithfully describe the measurable world.

Meaning begins only after mathematics has honestly completed its work.

The measurable and the meaningful therefore become complementary rather than competitive.

The Architecture of Meaning

The empirical sciences investigate mechanisms.

The Architecture asks another question.

What organizational significance does meaning introduce?

Awareness allows experience.

Learning organizes experience.

Memory preserves experience.
Knowledge organizes preserved experience.
Meaning now provides orientation.
Knowledge tells us what is true.
Meaning helps determine what deserves our care.
Meaning therefore transforms understanding into direction.
Reality remains the foundation.
Meaning becomes the compass.
The Architecture recognizes meaning as one of the principal organizers of responsible participation.

The Principle of Orientation

Every organized system eventually requires orientation.
Migration requires orientation.
Navigation requires orientation.
Scientific investigation requires orientation toward evidence.
Education requires orientation toward truth.
Civilization requires orientation toward justice.
Human life likewise requires orientation.
Meaning provides that orientation.
Meaning does not replace reality.
Meaning aligns participation with reality.
The Architecture therefore proposes another constitutional principle.
Meaning strengthens coherence when understanding increasingly guides faithful participation within reality.

Scientific Spiritualism

At this point another important distinction becomes necessary.
Scientific Spiritualism does not replace science.
It begins only after science has honestly spoken.
The Architecture therefore defines Scientific Spiritualism as disciplined reflection upon the interior dimensions of human existence in light of humanity's continually expanding scientific understanding of reality.
Questions of gratitude.
Responsibility.
Forgiveness.
Hope.

Beauty.
Wonder.
Purpose.
Compassion.
These belong neither exclusively to laboratories nor exclusively to religious traditions.
They belong wherever human beings seek to live wisely within reality.
Scientific Spiritualism therefore remains grounded.
Reality first.
Reflection second.
Science remains fully honored.

A Symbolic Recognition

The Architecture now introduces another symbolic organizational relationship.
Let
Me
represent meaningful orientation.
K
represent organized knowledge.
R
represent lived relationships.
P
represent responsible participation.
The Architecture symbolically recognizes:
$Me = f(K, R, P)$
Meaning deepens as organized knowledge increasingly guides faithful participation within reality.
This symbolic relationship is not proposed as an empirical psychological law.
It summarizes an organizational recognition repeatedly encountered across psychology,
anthropology,
history,
education,
and the lived human experience.
The empirical sciences investigate the mechanisms.
The Architecture recognizes the recurring organizational relationship they collectively illuminate.
Again,
the constitutional order remains unchanged.

Empirical foundation.
↓
Mathematical language.
↓
Architectural recognition.

The Spirit Field

The Spirit Field now enters the Architecture.

Not as another physical field.

Not as an additional force of nature.

Not as a competing scientific theory.

The Spirit Field is introduced as the Architecture's symbolic description of interior human coherence.

It names the lived integration of
understanding,
gratitude,
hope,
responsibility,
forgiveness,
wonder,
purpose,
and compassion.

Like every previous symbolic expression within this work,
the Spirit Field does not replace empirical science.

It provides a language for reflecting upon dimensions of human participation that science increasingly illuminates, yet was never designed to measure exhaustively.

Reality remains one.

Our participation within it becomes increasingly rich.

The Architect's Reflection

For many years I believed that meaning would appear after I found the right answers.

Reality quietly taught something different.

Meaning often appeared long before certainty.

It appeared while listening.

While serving.

While standing beside people whose dignity remained greater than their suffering.

Some of the deepest lessons of my life came not through success,

but through faithful presence.

Perhaps meaning is not discovered by escaping reality.

Perhaps meaning emerges through increasingly faithful participation within reality.

Quiet Reflection

Think of someone who quietly changed your life.

Perhaps they never became famous.

Perhaps they never published a book.

Perhaps they simply remained present when presence mattered most.

Ask yourself,

What gave that relationship meaning?

Remain with that question.

Meaning rarely arrives as abstraction.

Meaning usually arrives through relationship.

Crossing the Next Arch

Reality has now taught us much.

We have observed.

Measured.

Learned.

Remembered.

Organized.

Understood.

Found orientation.

One constitutional question remains.

Across every chapter,

one remarkable pattern has quietly repeated.

Reality continually preserves organized relationships despite continual change.

The Architecture has repeatedly recognized this pattern.

The next chapter finally gives it a name.

Coherence.

There,

for the first time,
the Cathedral Equation will appear.
Not as a new idea.
But as the mathematical recognition of a journey the reader has already completed.
The equation will not ask the reader to believe.
It will simply allow the reader to recognize what reality has been teaching since the first page.

Cathedral Thread

Constitutional Principle Established

Meaning is organized understanding that faithfully orients participation within reality.

Chapter One established:
Reality precedes explanation.
Chapter Two established:
Understanding deepens through relationship.
Chapter Three established:
Identity emerges through differentiation.
Chapter Four established:
Boundaries preserve identity.
Chapter Five established:
Vulnerability reveals value.
Chapter Six established:
Adaptation preserves identity through organized change.
Chapter Seven established:
Attention directs adaptive participation.
Chapter Eight established:
Awareness transforms attention into lived experience.
Chapter Nine established:
Learning reorganizes experience.
Chapter Ten established:
Memory preserves organized learning.
Chapter Eleven established:
Knowledge organizes preserved memory.
Chapter Twelve now establishes:
Meaning faithfully orients organized understanding toward responsible participation within reality.

The next chapter asks the constitutional question toward which the entire

Cathedral has been quietly moving:

What recurring organizational principle has every previous chapter been preparing us to recognize?

The Cathedral now opens its Keystone.

Chapter Thirteen

Coherence

The Keystone of the Cathedral

"Reality does not merely persist. It continually preserves organized relationships through change. The Architecture calls this coherence."

Guiding Question

What recurring organizational principle has every previous chapter been preparing us to recognize, and why does coherence become the keystone of the entire Architecture?

The previous twelve chapters have guided the reader through one continuous journey.

Reality exists.

Relationships organize reality.

Differentiation permits identity.

Boundaries preserve identity.

Vulnerability introduces challenge.

Adaptation preserves organization.

Attention selects.

Awareness experiences.

Learning reorganizes.

Memory preserves.

Knowledge organizes.

Meaning orients.

At every stage,

one remarkable observation has quietly repeated.

Organized systems continually preserve themselves despite continual change.

Stars endure.

Cells endure.

Forests endure.

Organisms endure.

Civilizations endure.

The mechanisms differ profoundly.
Yet one organizational pattern continually reappears.
This chapter finally gives that pattern a name.
Coherence.
The Architecture has not invented coherence.
It has gradually learned to recognize it.

Scientific Foundations

This chapter stands upon the inheritance of empirical science.

Its principal foundations include:

- Astrophysics
- Thermodynamics
- Systems Biology
- Physiology
- Neuroscience
- Ecology
- Information Theory
- Systems Science

Modern science does not define coherence with one universal meaning.
Instead,
each discipline encounters organized persistence through its own language.
Astrophysics investigates hydrostatic equilibrium within stars.
Physics investigates dynamically stable systems.
Chemistry investigates stable molecular organization.
Biology investigates homeostasis.
Cell biology investigates continual repair and molecular regulation.
Neuroscience investigates coordinated neural activity.
Ecology investigates resilient ecosystems.
Medicine investigates physiological regulation.
Systems science investigates organized interaction.
Information theory investigates reliable communication despite noise.
Each discipline faithfully investigates its own domain.

The Architecture receives every one with gratitude.

Only afterward does it ask what recurring organizational principle they collectively illuminate.

Mathematical Foundations

Mathematics likewise approaches organized persistence through many complementary languages.

Among them are:

- Differential Equations
- Control Theory
- Feedback Systems
- Information Theory
- Network Theory
- Graph Theory
- Topology
- Dynamical Systems

Each reveals one aspect of organized persistence.

None alone exhausts reality.

Together they illuminate an extraordinary observation.

Reality repeatedly preserves organization through continual interaction rather than static perfection.

The Architecture inherits these mathematical languages exactly where they belong.

They become the structural language through which coherence becomes increasingly intelligible.

Architectural Recognition

Only after empirical science and mathematics have spoken does the Architecture introduce its thirteenth constitutional recognition.

Coherence is the continued preservation of organized relationships sufficient for identity, function, adaptation, and meaningful participation within reality.

Coherence is not perfection.
It is not immobility.
It is not symmetry alone.
It is not equilibrium alone.
Coherence is organized persistence through continual change.
The Architecture recognizes coherence as the recurring organizational principle toward which every previous chapter has been quietly moving.

Coherence

The Keystone of the Cathedral

The previous twelve chapters have guided the reader through one continuous journey.

Reality exists.
Relationships organize reality.
Differentiation permits identity.
Boundaries preserve identity.
Vulnerability introduces challenge.
Adaptation preserves organization.
Attention selects.
Awareness experiences.
Learning reorganizes.
Memory preserves.
Knowledge organizes.
Meaning orients.
At every stage,
one remarkable observation has quietly repeated.
Organized systems continually preserve themselves despite continual change.
Stars endure.
Cells endure.
Forests endure.
Organisms endure.
Civilizations endure.
The mechanisms differ profoundly.
Yet one organizational pattern continually reappears.
The Architecture gives that recurring pattern one name.
Coherence.

The Mathematics of Organized Persistence

The scientific disciplines explain how coherent systems persist.

Mathematics provides the language through which those processes become increasingly intelligible.

Control theory investigates regulation.

Feedback theory investigates continual correction.

Differential equations describe evolving systems.

Network theory investigates resilient organization.

Information theory investigates meaningful communication despite noise.

Each mathematical language contributes another perspective.

Together they prepare the reader for one final architectural recognition.

The Cathedral Equation.

The Architecture of Coherence

The empirical sciences investigate the mechanisms through which organized systems persist.

The Architecture asks another question.

What recurring organizational principle becomes visible across these scientific disciplines?

The answer does not replace science.

It emerges from science.

Stars preserve organized structure through continual regulation.

Living cells preserve organization through continual repair.

Organisms preserve coordinated function through homeostasis.

Brains preserve meaningful activity through dynamic neural organization.

Ecosystems preserve resilience through continual adaptation.

Civilizations preserve continuity through education, memory, justice, and cooperation.

The mechanisms differ profoundly.

The recurring organizational observation remains.

Organized systems continually preserve relationships sufficient for continued participation within reality.

The Architecture recognizes this recurring principle as Coherence.

Coherence therefore is not perfection.

Not symmetry alone.

Not equilibrium alone.

Not immobility.

Coherence is the continued preservation of organized relationships sufficient for identity,

function,

adaptation,

and meaningful participation within reality.

This recognition becomes the Keystone of the Cathedral.

The Cathedral Equation

The reader has now encountered stars.

Cells.

Brains.

Immune systems.

Forests.

Civilizations.

Every chapter has quietly prepared one recognition.

The Architecture now gives symbolic expression to that recognition.

Let

C

represent the coherence of an organized system.

G

represent coherence-generating processes.

L

represent coherence-loss processes.

The Architecture symbolically recognizes:

$dC/dt = G - L$

This symbolic equation is not proposed as a new empirical law of physics.

It does not replace Einstein's Field Equations.

It does not replace Maxwell's Equations.

It does not replace Schrödinger's Equation.

It does not replace thermodynamics,

biology,

neuroscience,

or systems science.

Rather,

it summarizes one recurring organizational relationship repeatedly encountered across those scientific disciplines.

The mechanisms remain the responsibility of each empirical science.

The Architecture recognizes the recurring organizational pattern they collectively illuminate.

This constitutional distinction remains essential.

Empirical Foundation

↓

Mathematical Description

↓

Architectural Recognition

The Cathedral Equation belongs entirely within the third layer.

Reading the Cathedral Equation

Consider a star.

Gravity continually draws matter inward.

Thermonuclear fusion continually generates outward pressure.

The astrophysicist explains those mechanisms.

The Architecture asks another question.

Which processes preserve coherence?

Which processes diminish coherence?

Now consider a living cell.

Protein synthesis.

DNA repair.

Membrane regulation.

Energy metabolism.

Oxidative stress.

The cell biologist investigates each mechanism.

The Architecture recognizes their common organizational relationship.

Now consider civilization.

Education generates coherence.

Justice generates coherence.

Trust generates coherence.

Compassion generates coherence.

Violence,

ignorance,

corruption,

and fragmentation diminish coherence.

The historian investigates civilizations.

The Architecture recognizes another recurring pattern.

Different mechanisms.

One organizational grammar.

The Principle of Dynamic Coherence

The Architecture therefore proposes another constitutional principle.

Every organized system continually negotiates coherence through the interaction of coherence-generating and coherence-loss processes.

Persistence does not emerge because change ceases.

Persistence emerges because organized relationships remain sufficient despite continual change.

Reality demonstrates this principle repeatedly.

Stars.

Cells.

Brains.

Forests.

Civilizations.

Human lives.

The Architecture simply gives this recurring observation a common language.

The Geometry of Coherence

The reader may now recognize another remarkable progression.

Reality
↓
Relation
↓
Differentiation
↓
Boundary
↓
Vulnerability
↓
Adaptation
↓
Attention
↓
Awareness
↓
Learning
↓

Memory
↓
Knowledge
↓
Meaning
↓
Coherence
Nothing has been skipped.
Nothing has been forced.
Every chapter prepared the next.
The Architecture therefore reveals itself not as a theory imposed upon reality, but as a pathway discovered within reality.

Human Coherence

Human beings likewise negotiate coherence throughout life.

Health depends upon coordinated physiological organization.

Families depend upon trust.

Education depends upon disciplined learning.

Communities depend upon cooperation.

Civilizations depend upon justice.

The interior life depends upon integrity.

Coherence therefore extends beyond biology without abandoning biology.

Reality remains one.

The expressions of coherence become many.

The Spirit Field Revisited

The Spirit Field may now be understood more clearly.

It is not another physical field.

It is not another force.

It is not introduced to explain empirical mechanisms.

Rather,

it represents the Architecture's symbolic description of interior human coherence.

Gratitude.

Hope.

Forgiveness.

Purpose.

Wonder.
Responsibility.
Compassion.
These become interior expressions of coherent participation within reality.
Science illuminates the measurable.
The Spirit Field illuminates the lived.
Together they enlarge understanding without competing.

The Architect's Reflection

For many years, I believed I was searching for a new idea.

Reality patiently taught me that I was searching for an old pattern.

The pattern had always been there.

It appeared in the stars that maintained their brilliance through billions of years of ordered balance.

It appeared in the living cell that quietly repaired itself while the forces of entropy continually pressed against it.

It appeared in forests recovering after fire, in rivers finding their course, and in civilizations that endured because they preserved knowledge, justice, and compassion across generations.

Then I began seeing the same pattern in human lives.

I saw it in parents who quietly sacrificed for their children.

In teachers who patiently cultivated minds they would never fully see mature.

In physicians who stood beside suffering without surrendering hope.

In scientists whose willingness to be corrected became one of the greatest expressions of intellectual humility.

In ordinary people who chose integrity when no one else was watching.

Each was preserving something precious.

Each was participating in the same quiet architecture.

Gradually, I realized that I had not discovered coherence.

Reality had been demonstrating coherence all along.

The Architecture did not create the pattern.

It simply learned to recognize it.

That realization changed the purpose of this book.

It is not an argument for a new worldview.

It is an invitation to look more carefully at the world that has always been before us.

The chapters of this Cathedral have never asked the reader to abandon science.

They have asked the reader to honor it more deeply.

They have never asked the reader to replace mathematics.

They have asked the reader to appreciate mathematics as one of humanity's greatest languages for faithfully describing reality.

They have never asked the reader to diminish mystery.

They have asked the reader to distinguish mystery from confusion, and humility from uncertainty.

Coherence, then, is not something we impose upon reality.

It is something we gradually learn to recognize as we become more faithful students of reality itself.

Perhaps that is why the journey through this Cathedral begins with reality rather than belief, and ends not with certainty, but with responsibility.

For once we recognize that every enduring system depends upon preserving relationships worthy of continuation, we can no longer pretend that our own lives stand outside that pattern.

Every decision strengthens or weakens coherence.

Every act of honesty strengthens it.

Every act of compassion strengthens it.

Every faithful pursuit of truth strengthens it.

Every generation receives an inheritance shaped by the coherence of those who came before, and every generation leaves an inheritance shaped by its own choices.

That is why the Keystone is not merely an idea.

It is a responsibility.

The Cathedral does not ask us to admire coherence.

It asks us to participate in it.

And perhaps that has always been reality's quiet invitation.

Not merely to understand the world.

But to help preserve what is worthy within it.

Quiet Reflection

Before continuing, pause.

Do not think first about theories.

Do not think first about equations.

Do not think first about science.

Think instead about your own life.

Consider the people who have preserved your coherence.

Perhaps it was a parent who remained faithful through difficult years.

Perhaps it was a teacher who recognized possibilities you could not yet see in yourself.

Perhaps it was a physician, a friend, a colleague, or even a stranger whose kindness quietly changed the direction of your life.

Now consider the moments that tested your coherence.

Moments of uncertainty.

Loss.

Failure.

Illness.

Disappointment.

Unexpected change.

Ask yourself a simple question.

What helped you remain yourself while everything around you was changing?

Not what made life easy.

Not what made life successful.

What preserved your integrity?

Your hope?

Your compassion?

Your willingness to continue?

Remain with that question.

Across the pages of this Cathedral, coherence has never been presented as perfection.

It has never meant a life without struggle, uncertainty, or change.

It has meant something quieter.

The continued preservation of what is most worthy within us while reality continues to unfold around us.

Perhaps that same pattern has been present throughout your own life.

Not because every chapter unfolded according to your plans.

But because something essential continued to endure.

If you can recognize that quiet thread running through your own story, then you have already begun to understand the Keystone.

For coherence is not merely something we observe in stars, living cells, forests, or civilizations.

It is something we are invited to cultivate within our own lives.

Carry that thought with you as you enter the final chamber of the Cathedral.

For wisdom begins when understanding becomes the way we choose to live.

Crossing the Final Arch

The Cathedral has now reached its Keystone.

One movement remains.

Coherence alone is not the highest expression of organized reality.

Coherence prepares something greater.

Wisdom.

Compassion.
Stewardship.
Civilization.
The final chapter therefore asks one last constitutional question.
If reality continually teaches coherence,
what kind of civilization ought coherence produce?
That question completes the Cathedral.

Cathedral Thread

Constitutional Principle Established

Coherence is the continued preservation of organized relationships sufficient for identity, function, adaptation, and meaningful participation within reality.

Every previous chapter has quietly prepared this recognition.

The Cathedral Equation does not introduce a new truth.

It gives symbolic expression to one the reader has already recognized.

The final chapter now asks the Architecture's ultimate question.

What is the highest mature expression of coherent understanding?

The Cathedral now opens its final chamber.

Chapter Fourteen

Wisdom

The Highest Expression of Coherent Understanding

"Knowledge explains reality. Wisdom learns how to participate faithfully within it."

Guiding Question

What is the highest mature expression of coherent understanding, and how does wisdom transform knowledge into civilization?

Throughout this Architecture the reader has walked a long road.

Reality revealed itself through observation.

Science disciplined observation.

Mathematics revealed relationships.

Organization became increasingly visible.

Coherence gradually emerged.

None of these achievements alone, however,
guarantees wisdom.

Knowledge may increase.

Technology may increase.

Power may increase.

Information may increase.

Civilizations have repeatedly demonstrated that none of these alone ensures human flourishing.

Reality therefore asks another question.

How shall organized understanding become organized living?

This chapter explores the Architecture's final constitutional recognition.

Wisdom.

The Architecture recognizes wisdom not as greater intelligence,
but as coherent participation within reality.

Scientific Foundations

This chapter stands upon the inheritance of empirical science and the human sciences.

Its principal foundations include:

- Medicine
- Neuroscience
- Psychology
- Anthropology
- Sociology
- History
- Ethics

Science continually enlarges humanity's understanding of reality.

Medicine preserves life.

Engineering transforms civilization.

Physics reveals the universe.

Biology illuminates living systems.

Neuroscience deepens understanding of mind.

Every scientific discovery enlarges humanity's inheritance.

Science, however,

does not prescribe how every discovery ought to be used.

That responsibility belongs elsewhere.

Scientific understanding and human responsibility therefore become complementary.

The Architecture honors both.

Science reveals possibility.

Wisdom guides participation.

Mathematical Foundations

Mathematics continues serving the Architecture by faithfully organizing measurable relationships.

Systems theory.

Decision theory.
Game theory.
Network theory.
Optimization.
Probability.
These mathematical disciplines illuminate choices,
consequences,
cooperation,
and organization.
Yet mathematics itself does not determine what ought to be chosen.
It faithfully describes.
Wisdom faithfully orders.
The Architecture therefore receives mathematics with gratitude,
while recognizing that wisdom begins only after understanding seeks responsible participation within reality.

Architectural Recognition

Only after empirical science,
mathematics,
and organized knowledge have spoken does the Architecture introduce its final constitutional recognition.
Wisdom is coherent understanding faithfully lived.
Knowledge explains.
Meaning orients.
Wisdom participates.
The Architecture recognizes wisdom as the mature expression of coherent human becoming.

Wisdom

The Highest Expression of Coherent Understanding

Throughout this Architecture,
the reader has gradually discovered one remarkable truth.
Reality continually teaches.
Science continually discovers.
Mathematics continually describes.

The Architecture continually recognizes.
The question now becomes deeply personal.
How shall we live?
Wisdom begins when understanding becomes participation.
It is not measured by the quantity of information we possess.
It is measured by the quality of our participation within reality.

The Maturation of Coherence

The Architecture now reveals its complete constitutional progression.

Reality
↓
Observation
↓
Science
↓
Mathematics
↓
Organization
↓
Coherence
↓
Wisdom
↓
Compassion
↓
Stewardship
↓
Civilization

Nothing has been skipped.
Nothing has been assumed.
Everything has been inherited.
Everything has been earned.
The Cathedral now stands complete.

The Principle of Living Coherence

The Architecture therefore proposes its final constitutional principle.

Coherence reaches its highest expression when organized understanding con-

tinually strengthens life,
dignity,
responsibility,
and future inheritance.
Stars preserve cosmic organization.
Cells preserve biological organization.
Brains preserve cognitive organization.
Civilizations preserve cultural organization.
Human wisdom preserves moral organization.
Each level inherits those before it.
Each enlarges reality's continuing story.

Compassion

Wisdom naturally gives rise to compassion.

Not because compassion replaces science.

Because science continually reveals that every human being participates within one shared reality.

Every person inherits atoms forged within ancient stars.

Every person shares biological ancestry.

Every person experiences vulnerability.

Every person seeks meaning.

Every person eventually confronts mortality.

Compassion therefore emerges not despite science.

Compassion emerges because science continually reveals our profound interconnectedness.

The Architecture therefore recognizes compassion as one of coherence's highest mature expressions.

Understanding becomes relationship.

Relationship becomes care.

Care becomes civilization.

Stewardship

Reality entrusted every generation with an inheritance.

The Earth.

Knowledge.

Mathematics.

Language.

Culture.
Science.
Children.
None originated with us.
None belongs entirely to us.
Stewardship therefore becomes one of humanity's greatest responsibilities.

The Architecture understands stewardship as organized responsibility toward every inheritance entrusted to our care.

We inherit reality.
We preserve reality.
We enlarge humanity's understanding of reality.
Then we give it away.
Civilization continues.

Civilization

Civilization should never be measured solely by wealth.
Nor by technology.
Nor by military power.
Nor by monuments.

Civilization ultimately reveals itself through the quality of understanding faithfully transmitted across generations.

Civilizations flourish when education deepens.
When science remains honest.
When mathematics remains rigorous.
When justice remains trustworthy.
When compassion remains visible.
When wisdom continually reorganizes knowledge toward life.

The Architecture therefore recognizes civilization as organized coherence extending across generations.

Civilization is not merely built.
Civilization is continually taught.

The Living Stones

The Cathedral has never been built from paper.
Nor from equations.
Nor from ideas alone.
It has always been built from people.

Every teacher becomes another stone.
Every scientist becomes another stone.
Every physician becomes another stone.
Every engineer becomes another stone.
Every parent becomes another stone.
Every child who learns to love reality becomes another stone.
Books eventually age.
Buildings eventually weather.
Civilizations continually change.
The living stone remains.

Human beings faithfully shaped by reality become the enduring architecture through which civilization continues.

The Builder's Reflection

As these final pages came together, I gradually realized that this Cathedral was never truly about architecture.

Nor was it ultimately about science.

Nor mathematics.

Nor philosophy.

Each of these became faithful companions along the journey, but none was the destination.

The destination was always a way of seeing.

Reality had been teaching long before I possessed words to describe what I was seeing.

Science taught me to observe with discipline.

Mathematics taught me to appreciate the remarkable order hidden within relationships.

History taught me that every generation inherits more than it creates.

Medicine taught me that every life possesses immeasurable dignity.

Ordinary people taught me that wisdom often appears quietly—in perseverance, humility, sacrifice, forgiveness, curiosity, and hope.

My teachers became builders.

My parents became builders.

My patients became builders.

Scientists became builders.

Engineers became builders.

Children asking honest questions became builders.

Even those who challenged my thinking became builders, because they reminded me that reality is always greater than our explanations.

Gradually, I understood that no one builds Reality's Cathedral alone.

Every honest observation lays another stone.

Every mathematical insight strengthens another arch.

Every scientific discovery opens another chamber.

Every act of compassion becomes another window through which light enters.

The Cathedral has always belonged to reality.

We have simply been privileged to help reveal a little more of its architecture.

If this book leaves any lasting impression, I hope it is not that the reader remembers its author.

I hope the reader remembers to look more carefully at reality.

To ask better questions.

To welcome correction.

To love truth more than certainty.

To receive the inheritance of previous generations with gratitude.

And then to add one faithful stone before passing that inheritance forward.

That is how science grows.

That is how civilization matures.

That is how wisdom quietly enters history.

No Cathedral is ever truly finished.

Each generation strengthens the foundations it has inherited.

Each generation opens doors that previous generations could not yet see.

Each generation prepares rooms for people it will never meet.

This book is offered in that same spirit.

Not as the final word.

But as one carefully placed stone within humanity's continuing conversation with reality.

If, someday, another reader discovers a clearer pattern, a deeper mathematical relationship, a more faithful scientific understanding, or a wiser way of organizing what has been presented here, I hope they will improve it without hesitation.

Nothing would honor this work more.

Because the purpose of this Cathedral has never been to preserve its own permanence.

Its purpose has always been to encourage humanity to become increasingly faithful to reality.

And perhaps that is the quiet responsibility entrusted to every builder.

Not to construct monuments to ourselves.

But to leave the foundations stronger for those who follow.

If this Cathedral has helped you see reality a little more clearly...

Then its work has already begun to continue through you.

Quiet Reflection

Close the book for a moment.

Think about every person whose life quietly shaped yours.

Some were scientists.

Some were teachers.

Some were family.

Some were strangers.

Some never knew the influence they would have.

Ask yourself one final question.

What inheritance am I now preparing for those I will never meet?

Remain there.

Every generation receives foundations.

Every generation leaves foundations.

The measure of civilization is the integrity of the stones it chooses to lay.

The Final Inscription

Reality remains humanity's first teacher.

Science remains humanity's greatest discipline of discovery.

Mathematics remains humanity's most precise language of relationship.

Coherence remains humanity's discipline of organized understanding.

Wisdom remains humanity's faithful participation within reality.

Compassion remains humanity's highest mature expression of coherent living.

Every generation lays another stone.

The Cathedral remains open.

Always.

The Quiet Possibility

Throughout history,

humanity has often measured leadership through power.

Through conquest.

Through wealth.

Through technological achievement.

Reality quietly suggests another possibility.

As understanding deepens,

leadership itself may mature.

Not because power disappears.

Because responsibility grows.

The parent gradually learns that authority exists to protect.

The physician learns that knowledge exists to preserve life.

The scientist learns that discovery carries ethical responsibility.

The teacher learns that education shapes generations yet unborn.

Perhaps civilizations also mature in this way.

Perhaps one measure of an advancing civilization is not simply its capacity to accumulate knowledge,

but its capacity to organize that knowledge toward the flourishing of life.

If so,

then compassion is no longer sentiment.

Compassion becomes an expression of organized understanding.

The Architecture does not claim that history guarantees such a future.

Reality offers no such guarantee.

It simply recognizes a possibility.

As humanity increasingly understands the profound relationships connecting every living system,

leaders may gradually recognize that preserving human dignity strengthens civilization itself.

Such a future is neither inevitable nor impossible.

It remains a choice.

The Cathedral therefore offers no prediction.

Only an invitation.

Perhaps the future of civilization depends less upon the quantity of knowledge humanity possesses,

and more upon the wisdom with which that knowledge is lived.

If this possibility continues growing,

then every generation may inherit a civilization slightly more coherent,

slightly more compassionate,

and slightly more faithful to reality than the one before it.

That hope alone is worthy of careful stewardship.

Cathedral Thread

Final Constitutional Principle

Wisdom is coherent understanding faithfully lived.

The Cathedral has completed its journey.

Reality taught.

Science discovered.
Mathematics described.
The Architecture recognized.
Wisdom participated.
The Cathedral was never built merely to be admired.
It was built to be inhabited.

APPENDIX A

The Constitutional Order of Coherence Architecture

The constitutional rhythm of Coherence Architecture may be summarized as follows:

Reality
↓
Observation
↓
Empirical Science
↓
Mathematics
↓
Organization
↓
Coherence Architecture
↓
Reflection
↓
Wisdom
↓
Compassion
↓
Civilization

Readers now possess the complete map.

APPENDIX B

The Grand Architecture

The complete architecture may be viewed as one continuous progression:

Reality
↓
Relation
↓
Differentiation
↓
Boundary
↓
Vulnerability
↓
Adaptation
↓
Attention
↓
Awareness
↓
Learning
↓
Memory
↓
Knowledge
↓
Meaning
↓
Coherence
↓

Wisdom
↓
Compassion
↓
Stewardship
↓
Civilization
This progression presents the architecture of the entire book on a single page.

BIBLIOGRAPHY

By inheritance.

Foundations in Physics

Einstein
- Maxwell
- Planck
- Heisenberg
- Schrödinger
- Dirac
- Feynman
- Penrose
- Smolin

Mathematics

Euclid
- Hilbert
- Gödel
- Mandelbrot
- Shannon
- Von Neumann

Biology

Darwin
- Dobzhansky
- Mayr
- Margulis

Monod
Crick
Watson
Jacob
Nurse

Neuroscience & Cognition

Hebb
Edelman
Damasio
Tononi
Friston

Systems & Complexity

Bertalanffy
Ashby
Prigogine
Holland
Kauffman
Meadows

Philosophy

Aristotle
Whitehead
Peirce
Polanyi
Popper
Kuhn

Civilization & History

The monumental traditions of Nubia and Kemet
Classical Greek mathematics
The Scientific Revolution
Modern empirical science

APPENDIX C

Equation Registry

The Mathematical Architecture of Reality's Cathedral

A Note to the Reader

The symbolic equations presented throughout Coherence Architecture are not proposed as replacements for the empirical equations of physics, chemistry, biology, neuroscience, mathematics, or any other scientific discipline.

Rather, they serve a different purpose.

They summarize recurring organizational recognitions that emerge after empirical science has faithfully described reality.

For that reason, every equation in this Architecture follows the same constitutional order:

Empirical Foundation

↓

Mathematical Language

↓

Architectural Recognition

The equations therefore belong to the Architecture.

The empirical mechanisms remain the responsibility of the scientific disciplines from which they were inherited.

Equation I

Attention

Empirical Foundation

Neuroscience demonstrates that adaptive systems continually distinguish relevant information from competing stimuli.

Information theory distinguishes meaningful signal from surrounding noise.

Behavioral biology demonstrates continual selective attention across living systems.

Mathematical Language

Information Theory
Signal Detection Theory
Probability
Network Theory

Architectural Recognition

$A \propto R / N$

Where:
A = Effective Attention
R = Relevant Information
N = Competing Informational Noise

As relevant information becomes increasingly distinguished from competing noise,
adaptive participation generally improves.

Equation II

Awareness

Empirical Foundation

Neuroscience,
comparative cognition,
predictive processing,
memory integration.

Mathematical Language

Network Theory
Information Theory
Predictive Modeling
Systems Theory

Architectural Recognition

W = f(A, L, M)
Where:
W = Organized Awareness
A = Attention
L = Learning
M = Memory
Awareness deepens through increasing integration of attention, learning,
and memory.

Equation III

Learning

Empirical Foundation

Neuroplasticity
Educational Psychology
Bayesian Updating
Evolutionary Learning

Mathematical Language

Probability
Bayesian Inference
Recursive Mathematics
Learning Algorithms

Architectural Recognition

L = f(E, A, R)
Where:
L = Organized Learning
E = Experience
A = Attention
R = Reflection
Learning reorganizes experience into future participation.

Equation IV

Memory

Empirical Foundation

Genetics
Neuroscience

Immunological Memory
Education
Information Preservation

Mathematical Language

Information Theory
Coding Theory
Graph Theory
Recursive Systems

Architectural Recognition

$M = f(L, I, P)$
Where:
M = Organized Memory
L = Learning
I = Inherited Organization
P = Preserved Patterns
Memory allows the past to continue participating in the future.

Equation V

Knowledge

Empirical Foundation

Scientific Organization
Pattern Recognition
Theory Formation
Evidence Integration

Mathematical Language

Graph Theory
Information Theory
Linear Algebra

Systems Theory

Architectural Recognition

K = f(M, R, I)
Where:
K = Organized Knowledge
M = Memory
R = Recognized Relationships
I = Integrated Understanding
Knowledge organizes preserved memory into increasingly faithful understanding.

Equation VI

Meaning

Empirical Foundation

Psychology
Anthropology
Neuroscience
History
Human Experience

Mathematical Language

Decision Theory
Network Theory
Systems Theory

Architectural Recognition

Me = f(K, R, P)
Where:
Me = Meaning
K = Knowledge

R = Relationships
P = Participation
Meaning orients organized understanding toward faithful participation within reality.

Equation VII

The Cathedral Equation

Empirical Foundation

Hydrostatic Equilibrium
Homeostasis
Adaptive Immunity
Ecological Resilience
Neural Regulation
Civilizational Continuity

Mathematical Language

Differential Equations
Control Theory
Feedback Systems
Network Theory
Information Theory
Dynamical Systems

Architectural Recognition

$dC/dt = G - L$
Where:
C = Coherence
G = Coherence-Generating Processes
L = Coherence-Loss Processes
The Cathedral Equation summarizes one recurring organizational relationship repeatedly encountered across many scientific disciplines.
It is not proposed as a new empirical law.

It is an architectural recognition of organizational persistence.

Looking Forward

The equations presented in this volume are intentionally introductory.

Their purpose is educational rather than formal.

They summarize organizational recognitions that arise naturally as the reader progresses through Reality's Cathedral.

The complete mathematical development,

formal definitions,

relationships among the symbolic equations,

and their deeper architectural implications are developed in the companion volume:

The Mathematics of Becoming

There,

the Architecture moves from recognition to formal exploration.

Book I teaches the reader how to see.

Book II teaches the reader how to work with what has been seen.

APPENDIX D

The Great Conversation

Those Who Helped Humanity Read Reality

Before the Reader Begins

Reality has never belonged to one civilization, one discipline, one generation, or one individual.

Human understanding has always advanced through conversation.

Observation became measurement.

Measurement became mathematics.

Mathematics became prediction.

Prediction became deeper understanding.

Each generation inherited discoveries from those who came before and added its own contribution before passing that inheritance forward.

This appendix is therefore not presented merely as a bibliography.

It is offered as The Great Conversation.

The individuals and traditions listed here did not all agree with one another.

Nor did they answer the same questions.

Together, however, they enlarged humanity's ability to read reality more faithfully.

Coherence Architecture gratefully receives their contributions.

Foundations in Mathematics

The Architecture honors those whose work provided humanity with increasingly precise languages for describing reality.

- Euclid — Geometry
- Archimedes — Mathematical Physics
- René Descartes — Analytic Geometry
- Isaac Newton — Calculus (co-development)
- Gottfried Wilhelm Leibniz — Calculus (co-development)
- Carl Friedrich Gauss — Number Theory, Geometry
- David Hilbert — Foundations of Mathematics
- Kurt Gödel — Mathematical Logic
- John von Neumann — Linear Algebra, Computing, Systems
- Claude Shannon — Information Theory
- Benoit Mandelbrot — Fractal Geometry

Foundations in Physics

The Architecture honors those who enlarged humanity's understanding of matter, energy, motion, and spacetime.

- Galileo Galilei
- Isaac Newton
- Michael Faraday
- James Clerk Maxwell
- Max Planck
- Albert Einstein
- Niels Bohr
- Werner Heisenberg

- Erwin Schrödinger
- Paul Dirac
- Richard Feynman
- Subrahmanyan Chandrasekhar
- Roger Penrose
- Lee Smolin

Foundations in Chemistry

The Architecture honors those who revealed the organization of matter.

- Antoine Lavoisier
- John Dalton
- Dmitri Mendeleev
- Gilbert N. Lewis
- Linus Pauling

Foundations in Biology

The Architecture honors those who enlarged humanity's understanding of life.

- Charles Darwin
- Alfred Russel Wallace
- Gregor Mendel
- Theodosius Dobzhansky
- Ernst Mayr
- Lynn Margulis
- Jacques Monod

- François Jacob
- Francis Crick
- Sydney Brenner
- Paul Nurse

Foundations in Medicine

The Architecture gratefully acknowledges the physicians, nurses, researchers, and caregivers whose work continues preserving human life.

Representative contributors include:

- Hippocrates
- Andreas Vesalius
- William Harvey
- Florence Nightingale
- Ignaz Semmelweis
- Louis Pasteur
- Robert Koch
- Alexander Fleming

The Architecture also honors the countless unnamed clinicians, researchers, caregivers, and patients whose lives continue expanding humanity's understanding of health, disease, healing, and dignity.

Foundations in Neuroscience and Cognition

The Architecture receives with gratitude the work of those who enlarged humanity's understanding of mind and learning.

- Donald Hebb
- Gerald Edelman

- Antonio Damasio
- Daniel Kahneman
- Michael Posner
- Giulio Tononi
- Karl Friston

Foundations in Systems and Complexity

The Architecture particularly acknowledges those whose work made organizational thinking increasingly possible.

- Ludwig von Bertalanffy
- W. Ross Ashby
- Norbert Wiener
- Ilya Prigogine
- John Holland
- Stuart Kauffman
- Donella Meadows

Foundations in Philosophy of Science

The Architecture also stands within a long tradition of disciplined reflection concerning knowledge itself.

- Aristotle
- Francis Bacon
- Charles Sanders Peirce
- Alfred North Whitehead

- Michael Polanyi
- Karl Popper
- Thomas Kuhn

Foundations in Civilization

Reality has also been faithfully observed long before modern scientific institutions.

The Architecture therefore gratefully acknowledges the cumulative contributions of civilizations that cultivated mathematics, astronomy, medicine, engineering, ethics, and systems of knowledge.

Among them are:

- The scientific and mathematical traditions of Ancient Kemet
- The monumental and astronomical traditions of Nubia and Kush
- Mesopotamian astronomy and mathematics
- Classical Greek philosophy and mathematics
- Indian mathematical traditions
- Chinese scientific traditions
- Islamic Golden Age scholarship
- The European Scientific Revolution
- Modern global empirical science

The Architecture recognizes that humanity's understanding of reality has always been cumulative.

Every civilization has contributed another stone.

The Great Conversation Continues

The names listed here are neither exhaustive nor final.

Many of the greatest builders of Reality's Cathedral remain unknown.

Teachers who never published.

Parents who quietly shaped generations.
Physicians who comforted the dying.
Engineers who solved forgotten problems.
Students whose questions opened new discoveries.
Caregivers whose compassion preserved dignity.
Every honest seeker enlarges humanity's inheritance.
Reality continues teaching.
Science continues discovering.
Mathematics continues revealing relationships.
The Great Conversation therefore remains unfinished.
May future generations enlarge it with humility.

Closing Reflection

Reality's Cathedral belongs to no author.

It belongs to Reality.

Every generation enters through a different doorway.

Every generation leaves another stone.

May those who read this work continue the conversation with honesty, courage, gratitude, and humility.

MAPS OF REALITY'S CATHEDRAL

Visual Architecture of Coherence

"The following maps are not intended to replace the chapters that precede them. They reveal, in visual form, the architectural journey the reader has already completed. They are offered as navigational aids through Reality's Cathedral."

MAP I

Reality's Cathedral

The Grand Journey

REALITY
↓
RELATION
↓
DIFFERENTIATION
↓
BOUNDARY
↓
VULNERABILITY
↓
ADAPTATION
↓
ATTENTION
↓

AWARENESS
↓
LEARNING
↓
MEMORY
↓
KNOWLEDGE
↓
MEANING
↓
COHERENCE
↓
WISDOM
↓
COMPASSION
↓
STEWARDSHIP
↓
CIVILIZATION
Caption:
The constitutional journey through Reality's Cathedral. Each stage inherits every stage before it.

MAP II

The Constitutional Order

How the Cathedral Thinks

Reality
↓
Observation
↓
Empirical Science
↓
Mathematics
↓

Organizational Recognition
↓
Reflection
↓
Wisdom
↓
Participation
Caption:
Every chapter follows this constitutional order. Nothing is recognized before it has first been observed.

MAP III

The Three Eternal Lights

Reality
▲
|
Science ◄——+——► Mathematics
|
▼
Coherence Architecture
Caption:
Reality remains the source. Empirical Science discovers. Mathematics describes. The Architecture recognizes.

MAP IV

Reality's Cathedral

The Great Builders

Reality
↓
Physics

Chemistry
Biology
Medicine
Neuroscience
Mathematics
Systems Science
Humanities
↓
Coherence Architecture
Caption:
Every discipline contributes another room to Reality's Cathedral.

MAP V

The Story of Becoming

Reality
↓
Matter
↓
Life
↓
Adaptation
↓
Attention
↓
Awareness
↓
Learning
↓
Memory
↓
Knowledge
↓
Meaning
↓
Coherence
↓

Wisdom
↓
Civilization
Caption:
The progressive maturation of organized participation within reality.

MAP VI

The Equation Journey

Attention
$A \propto R/N$
↓
Awareness
$W = f(A,L,M)$
↓
Learning
$L = f(E,A,R)$
↓
Memory
$M = f(L,I,P)$
↓
Knowledge
$K = f(M,R,I)$
↓
Meaning
$Me = f(K,R,P)$
↓
Coherence
$dC/dt = G - L$
Caption:

The symbolic equations follow the same order as the Cathedral itself. Every equation is earned before it appears.

MAP VII

Interior and Exterior

Exterior Reality
↓
Observation
↓
Science
↓
Mathematics
↓
Architecture
↓
Interior Participation
↓
Meaning
↓
Wisdom
↓
Compassion
↓
Stewardship
Caption:
Reality is encountered externally and lived internally. The Architecture recognizes the bridge between the measurable and the meaningful.

MAP VIII

The Cathedral Equation

Scientific Disciplines
↓
Physics
↓

Biology
↓
Neuroscience
↓
Ecology
↓
Systems Science
↓
Recurring Organizational Pattern
↓
$dC/dt = G - L$
Caption:

The Cathedral Equation does not replace empirical science. It summarizes one recurring organizational relationship revealed across many scientific disciplines.

MAP IX

The Educational Journey

Reality teaches
↓
Science discovers
↓
Mathematics describes
↓
Architecture recognizes
↓
Wisdom participates
Caption:
The constitutional rhythm of the entire book.

MAP X

The Living Stone

Reality
↓
Understanding
↓
Character
↓
Responsibility
↓
Stewardship
↓
Another Builder
Caption:
The Cathedral ultimately continues not through books alone, but through people who faithfully participate within reality.

MAP XI

The Great Conversation

Ancient Civilizations
↓
Classical Inquiry
↓
Scientific Revolution
↓
Modern Science
↓
Future Generations
Caption:
Reality's Cathedral has never belonged to one civilization or one generation. Humanity's understanding continually expands through the Great Conversation.

MAP XII

The Cathedral Completed

FOUNDATION
Reality
Science
Mathematics
↓
PILLARS
Relation
Differentiation
Boundary
↓
ARCHES
Vulnerability
Adaptation
Attention
Awareness
↓
KEYSTONE
Coherence
↓
LIVING STONES
Wisdom
Compassion
Stewardship
Civilization
Caption:
The complete constitutional architecture of Reality's Cathedral.

ABOUT THE AUTHOR

Aston Gustavous Farquharson is the founder of the NuSpecies Integrative Practice and the originator of the Singularity Dual-Aspect Emergence (SDAE) framework, a multidisciplinary body of work exploring the relationships among cosmology, mathematics, biology, consciousness, and human flourishing.

Over several decades, his work has been shaped by scientific study, engineering, natural health practice, and thousands of conversations with individuals and families seeking not only better health, but a deeper understanding of reality itself. These experiences led to the development of Coherence Architecture, an educational framework that organizes empirical science, mathematics, systems thinking, and lived human experience into a coherent pathway of understanding while preserving the integrity of each discipline.

Rather than presenting a new scientific theory, Coherence Architecture serves as the constitutional framework through which the broader SDAE body of work may be understood. It begins with empirical reality, follows the discoveries of modern science and mathematics, recognizes recurring organizational patterns across disciplines, and ultimately asks how humanity may live wisely within the reality it continues to discover.

His work is guided by one enduring conviction:

Reality has always been my first teacher.

A Living Architecture

A Closing Note to Future Readers

This First Foundational Edition represents the best understanding available at the time of its publication.

It is offered with gratitude to the countless generations whose observations, questions, discoveries, and sacrifices have enlarged humanity's understanding of reality.

The Architecture presented throughout these pages does not claim finality.

It claims responsibility.

Responsibility to begin with reality.

Responsibility to honor empirical science.

Responsibility to respect mathematics as humanity's most precise language for describing measurable relationships.

Responsibility to distinguish observation from interpretation, and explanation from recognition.

Reality continues to teach.

Science continues to discover.

Mathematics continues to reveal new relationships.

Human understanding continues to mature.

For these reasons, this Architecture is intentionally living rather than static.

Future discoveries should strengthen it.

Better mathematics should refine it.

Deeper scientific understanding should illuminate it.

Clearer organizational recognition should expand it.

Every correction honestly earned through greater faithfulness to reality is welcomed, not resisted.

No generation possesses the final word.

Every generation inherits.

Every generation learns.

Every generation contributes.

Every generation leaves another stone for those who follow.

If future readers discover clearer patterns than those presented here, may they preserve what remains faithful, improve what can be strengthened, and continue the Great Conversation with the same humility that has guided humanity's greatest scientific and intellectual achievements.

The measure of this work will never be whether every sentence remained unchanged.

Its measure will be whether it continued encouraging people to become more faithful students of reality.

For Reality remains humanity's first teacher.

May this Cathedral remain open.

May its foundations remain strong.

May its builders remain humble.

And may those who enter after us leave it stronger than they found it.

www.ingramcontent.com/pod-product-compliance
Lightning Source LLC
LaVergne TN
LVHW050930210926
848482LV00013B/596

* 9 7 9 8 9 9 6 6 8 0 5 2 8 *